MW00466254

KINGS OF A LONELY KINGDOM

EARTH DAY ESSAYS, POEMS, AND MUSINGS ON NATURE

DAVID C. MAHOOD

KINGS OF A LONELY KINGDOM

Copyright © 2021 by David C. Mahood

All rights reserved. This book or any portion thereof may not be reproduced or used in any manner whatsoever without the express written permission of the publisher except for the use of brief quotations in a book review.

First Printing, 2021

ISBN 978-0-9994876-2-4

Book design by Gina Poirier, Gina Poirier Design
front cover art: butterfly, planet earth: istockphoto.com

Olive Designs, LLC
191 E. Lothrop St.
Beverly, MA 01915

To Alex, Christopher, Katie, and Sam,

my sons and stepdaughters.

Live fully and tread lightly,

that our past foibles not be repeated.

CONTENTS

ACKNOWLEDGMENTS

⌇

Anyone who has written a book knows too well that it takes a group of dedicated people to get it to completion. I've once again leaned on a number of them. To start, I must thank my faithful first reader, Barb McGrath, who luckily also happens to be family, for her critical review. My copy editor par excellence, Cindy Black, who seems to make sense of what I draft. My talented book designer, Gina Poirier, who has an eye for the artistic. To Meg and Julie of Copper Dog Books—a big thanks for proving that a local indie bookstore, and not that online behemoth, can be an author's best friend. Likewise, to former bookstore owner Wendy Alexander-Persse, who keenly knows the struggle of new authors. To the New Jersey McGraths, too many to single out, thank you for sharing your stories. Thanks go also to my highly literate neighbors, the Balfs and the Harrops, who keep me informed and engaged. Sound editorial suggestions were also given by Charlotte Brummett. A shout out to Paul and Robin Satryb, eco-minded pals, who are ample sources of help and support. My friends, Susan Inglis and Angie Kenny, of Sustainable Furnishings Council, who are forces of good for nature. To Shelia Lohmiller and other talented friends from NEWH, thank you for giving me a voice within the network. I received some green building industry expertise from childhood friend, Tim Hughes. I always appreciate the help of my young eco-warrior friend, Rebecca Marasco, who gathered valuable feedback for me. Thanks also to fraternity brother and college professor, Mark Westneat, who really is a marine biologist, for his review and edits. And to another Wooster classmate and educator, Gretchen Marks Crane, for pitching in at the end.

To my mother, Bobbi Mahood, an avid reader and reviewer, who is so well read and loved. My lone sibling, Bruce Mahood, who made it into this manuscript, albeit briefly, is a fountain of encouragement. Finally, to my wife, Maryann, and our combined brood of four, Alex, Christopher, Katie, and Sam, you all are the sunshine of my life.

INTRODUCTION

⤳

I 've been writing about climate change, biodiversity and habitat loss, and other environment-related issues for close to twenty years. Additionally, I've been writing a series of Earth Day essays for a decade. I empathize with others who lament the loss of nature in a variety of forms. Most of my sentiments cover the topic of sustainability as an outshoot of my business interests in sustainable practices and products. But it all meets somewhere in the middle. Our world is out of ecological balance, and so many converging themes portend a looming crisis. The past ten years, every twenty-second of April—Earth Day—I have selected one issue to expose this and initiate a more serious reflection on our times, each about how far we have veered off track and about serious impacts on a sustainable future for all species. Throughout these chapters are the underlying themes of environmental pollution and climate change, corporate greed and accountability, injustice and activism, and endangered species. Within each chapter are hopeful solutions and even, at times, desperate examples of the actions of others. Six years ago, I began my first book with the word *inspiration* in the title. In this one, although we still need sources of inspiration, I am making the case that we can't wait for them to get involved. We've lingered too long in the waters of inaction, and the voices that most need to be heard are the ones rendered mute.

I was not formally trained in the sciences, more in literature and the languages. When I took a deeper interest in climate science, I was well into my thirties. But matters of ecology and the environment shadowed me from early on. I recently rediscovered a booklet I wrote in fourth grade that expressed this concern for life on Earth, complete with hyperbole and the childlike ability to emote without borders—damn, I miss that literary freedom. By the time I pursued a green MBA, I had published a number of articles on environment-related topics. By the completion of that degree, I had an outline of the meandering course I had followed to get there. The message I attempted to convey throughout the book was that it doesn't matter how or why one becomes environmentally active. The if and when do matter greatly, however. And now more than ever.

This book is not as scripted as my first book, *One Green Deed Spawns Another: Tales of Inspiration on the Quest for Sustainability*. In addition to the Earth Day essays are a combination of my poems and snippets of thoughts related to the environment over the past two decades or more. The selected themes were shaped by events of the year and my growing concern for the slow pace of action to combat global climate change and its interrelated complications—the many nasty tentacles. I have attempted to put each of these in some current context with related thoughts in equal parts, angst, care, disdain, and hope. Some of my first memories of expressing empathy for other species were likely a result of watching episodes of *Mutual of Omaha's Wild Kingdom* or *The Undersea World of Jacques Cousteau*, or maybe it was the film *Born Free*, introducing me to Elsa the Lioness. It could have been growing up with oversized black labs or one of my grandpa's barn cats. The origin of my empathy or biophilia, loosely defined as an inherent love or appreciation for fellow species, is probably undefined. Considering that close to half of all Americans share homes with dogs and cats, I guess I'm not alone in my feelings. Somehow it didn't recede in me over time. As I began to comprehend the delicate balance that exists in nature, the feelings only intensified. We have yet to map out our home, our living ecosystem, for its bountiful membership in this elite experience of life. Habitat destruction is as cruel as it is finite. That the web of life is now woven by one host species is beneath our capacity for community and connection, the loss of which is irretrievable even if we knew how to look for it, proving to be a more recent flaw of humankind. It occurred to me that from my earliest days, I had a curiosity about the world and its endless storybook. I hardly knew fifty years or more later I'd be running out of chapters.

CHAPTER

What's Goin' On?

If we all make an effort to understand the myriad of consequences of not only climate change but anthropogenic activity in general, we can't possibly turn our backs on them. We've already begun to change the world that our children live in and the one that they'll pass on to their children. For a people who embalm dead bodies for posterity, we don't preserve nature very well. She, our living planet, who has lately been regurgitating a lot in the form of hurricanes, wildfires, and tornadoes, has also been wheezing for decades. Included in this is a reshaping of the landscape and a reshuffling of the biota that survive. When I think of the awe-inspiring traits of other species, I can't imagine anything short of a global call to preserve them. Who would mindfully kill off a rare, sentient, living, breathing creature? Moreover, why would anyone hunt a species to extinction? What culture would sanction the slaughter of large mammals? No one in any developed country can make the argument that we are simply following our genetic instincts to hunt other creatures for the survival of the tribe. We don't kill other species because we need to eat them to stave off starvation or use their skin or hair for warmth or shelter; we extinguish them because they're in the way, and knowingly or unwittingly we're doing it all the same. Our explosive rise to the top of the evolution pyramid does not equate to free reign because as my biologist friend Mark Westneat states, "In the tree of life we are but one twig." Big brains and evolutionary and technological advancements have not proven to be very hospitable to other species as I described in a poem years ago. Gaia represents a life-giving being, a living Mother Earth, in this context.

By the Grace of Gaia

We sharpen sticks in dreams,
chase down ten-point bucks,
fast asleep under acrylic blankets.
Now upright and indoors,
We're shaded at work
inside cubicles,
confronting nature
through screensavers.

So why can't we rest
in treetops with howlers
or plunge like penguins?
Why can't we navigate
like sea turtles or
hear like horseshoe bats?

We can't weave orb spider webs
and we build jet engines
to plummet like falcons,
and nuclear-powered
submarines to dive
like sperm whales.

But instead of awe,

cohabitation,

we pinned tusks to walls

plumed our hats, festooned

our bodies with furs,

raided the last gift

She anticipated

for Her progeny.

We have misunderstood

the real meaning of

evolution, and

we reign unopposed

by unnatural

selection and

survival of the unwise.

In the next chapter,

we learn loneliness.

(2006)

It is estimated that a third to half of all other species that inhabit Earth are at risk of extinction. Granted, most are casualties of habitat reduction and a lack of adaptability. Others will be displaced by invasive species—some of which will thrive in a warmer climate. Stripping our world of biodiversity is like removing the colors of the rainbow. Imagine if every future natural experience was in grayscale? At first it would look cool, like a disco ball in a nightclub, but after the third or fourth rotation all you would want to do is pummel it like a piñata. The variety of life on our planet is that sensory smorgasbord we take for granted. We'll lament the loss of species only when they are gone from view. It's been that way for most everything. I ascribe my feelings to biophilia, which I commented on a number of years ago.

> I am now, what I didn't self-diagnose before, a biophiliac.
> There is neither medication nor treatment for my illness.
> I have an innate love for our fellow species, which is the
> main symptom of my illness.

I began, also, to write Earth Day essays in 2010 to make a subtle point about our lack of urgency in arresting climate change and the loss of biodiversity. These essays share a common theme: What role should we take as Homo sapiens? Are we banishing our favorite species to theme parks? Mostly I believe we need to rethink our mission. By continuing our path of environmental plunder, we are skimming off many other species that cohabit this earth. *Homo sapiens*, Latin for *wise man*, is sometimes neither wise nor caring. One of my favorite environment writers, Elizabeth Kolbert, has put it more succinctly in her monumentally important book, *The Sixth Extinction: An Unnatural History*: "Having been alerted to the ways in which we're imperiling other species, can't we take action to protect them? Isn't the whole point of trying to peer into the future so that, seeing dangers ahead, we can change course to avoid them?"[1] We are delinquent, even criminal in some cases, if we bargain one more native species for the extraction of natural resources like fossil fuels. We're dooming some simply as a result of a century or more of burning coal, oil, and gas. Let's hope polar bears live on forever in Coca-Cola commercials because they certainly aren't gonna here. And you'll only find Nemo on TV in the next twenty-five years because clownfish like other reef fish are very susceptible to climate change. Instead of learning kung fu, pandas need to learn how to breed more often.

And what rights should be granted to other species? Who owns the rights to nature anyway? We humans haven't exactly demonstrated that we should own them. So, in effect, we are grappling with ethical issues when we address anthropogenic climate change and biodiversity. I have been accused of caring more about other species than our own but that fundamentally misses the point. Simply because we evolved beyond the great apes and Neanderthals doesn't grant us the right to misuse our habitat or endanger its diversity of life. We haven't even come close to identifying all of the species that inhabit this planet, and we're on a course to extinguish a third of them.[2]

I care about other species because I care so much about our own. Every parent sets out to provide a safe and loving home for their children, but if we have diminished the experience of future generations to enjoy the natural gifts of this planet then we have failed them. And, yes, kids, you have a right to be angry, and to protest, and to confront your political leaders. Although the predictions may ultimately happen, the outcome shouldn't have been predestined. The efforts of so many committed environmentalists, and everyday citizens, prove to me that the will exists. I profiled a dozen of them in *One Green Deed Spawns Another*.

The challenges ahead are many, all of them of great significance. And I am not absolving myself of my role in the state of this planet but I have been expressing concern for the health of it for a long time. I am not an oblivious occupant of Earth, however; I can truly make that distinction. If I have grandchildren, I am aware that they will not have the same sights and experiences in nature that I have had. Food, water, and shelter will take on a whole new meaning in their lifetimes. I am deeply troubled by the sacrifices future generations may have to make as a result of the ones our generation failed to make.

Why We Care

Climate news of the twenty-first century has become first page news unlike the century before, and that is a good and bad scenario. It truly sucks that it is a frequent, dour event but it is also good that people spanning the globe are reacting to the same news. Say what you want about social media but if you haven't heard of the numerous global movements confronting climate change, species extinction, plastic and water pollution, mining, blasting, and messing with land, water, and air, in general, you may want to check your computer cache or visit your kids' schools on Fridays. This movement is brimming with energy from our youngest to our oldest, and the rest of us in between. It is akin to the first Earth Day. Our children aren't sitting on the sidelines for this one either. Some of them, including Greta Thunberg, have become international figures.[3] Sunrise Movement and Extinction Rebellion, to name two organizations, are consistently confronting politicians around the world, further supporting the premise that people can initiate great change.[4] We are, at its base, irresponsible if we truly harm their future. Treading water while action was required makes me (us) complicit. Youthful energy motivated millions on April 22, 1970, as it is doing again today—albeit with higher stakes. I'm not only a part of it, I am in awe of the enormity of it all. You wanna see your life pass by you, pay no heed to the environmental protests of today.

In light of all that awareness, current weather crises related to climate change have become so common that I still want to scream out like a drunken town crier, "California wildfire kills 86 people.[5] Hurricane Irma devours Virgin Islands as a category 5 hurricane.[6] It's 45°F at the North Pole in the winter—yes, in the winter."[7] For someone who has been paying attention to these reports for quite some time, it can be overwhelming. And, periodically, I do cry out in protest, but I have also decided that thoughtful dialogue is in short supply. There are a bevy of town criers out there. These days we have enough of them—far more effective ones than I have been, and some who should just listen and be quiet. Mother Earth needs a bullhorn and her legions of followers do need to speak out but hyperbole and rhetoric are not helpful to the cause. I'm not likely to go silent but problems also require solutions, and we have to dedicate time to both: understanding the causes of our environmental crises and implementing the measures to solve them. The good news is that I am joined by millions of other citizens of the planet seeking the same approach.

I have chosen to make sense of all this through my musings and my April entreaties. Mostly I have tried to provide a voice for those that cannot speak out or for those whose voices have been muzzled. So maybe I still am a town crier, just more for the hairy, howling, blooming, pupating, pelagic, encroached upon, imperiled, incarcerated, and indigenous types. I want Earth Day to have deeper meaning and represent more than a twenty-four-hour period. Yes, sure, it is a good day to plant trees and pick up trash but our dependence on nature far exceeds that. If we can form these sinuous strands of environmental preservation into actions, then we have the makings of a new Earth Day movement. One more like its initial purpose.

The United States of America supersized in a period of about 250 years, and our economy became the largest in the world. We also transformed from an agrarian society to an industrial one. The Industrial Revolution in the United States coincides with the beginning of our assault on the atmosphere. Unregulated emissions along with natural occurrences began to alter the amount of greenhouse gases being trapped in our skies. In the United States, we started testing for global warming as a result of greenhouse gases in the atmosphere as early as the 1950s. Still, no U.S. president since World War II, Democrat or Republican, has been truly successful in putting together effective enough legislation to mitigate climate change. To most of our dismay, environmental issues have seeped into the world of politics, which has effectively silenced many scientists. Who could blame them? In 2020, it is the GOP, but irrespective of that, climate change is a bipartisan problem that both political parties can claim and must claim.

I've been out there for twenty years, and I am marched out— well, my knees and hip certainly are marched out. We haven't learned our lessons well. We've seen what industry, chemicals, and negligence looks like, and the worst is yet to come. By now most of us know that climate change is impacting our world through melting ice and permafrost, rising sea levels and emissions, dying coral reefs and pollinators, stronger storms and consequences, compromised habitats and species, migrating pests and environment refugees, altered ocean chemistry and air quality, and finally, engaged youth.

Our children are faced with the challenges of a century of blatant environmental damage. The modern environmental movement— post-World War II—began as a result of embittered youth also. Fed up with the lack of truth about the Vietnam War, the potential of being drafted, racial inequality, the search for a more free-spirited lifestyle from the prior decade, and the exposure of the chemistry industry for lying to the public about the dangers of readily used chemicals like DDT (dichlorodiphenyltrichloroethane) caused some of the unrest of the late 1960s and early 1970s. I was born into that time. But now that the Intergovernmental Panel on Climate Change has indicated that we have less time to gamble with atmospheric composition than we thought, it's no longer a tether ball for politicians to bat back and forth.[8]

Flaws come in a lot of shapes. I can forgive errors of judgment. Honest errors of judgment occur frequently, and some do have lasting environmental consequences. The errors caused by the technicians at Chernobyl nuclear power plant, for example, weren't done intentionally but they proved to be as tragic as if they were. The ones that are unforgivable tend to be connected to greed. Profiting at the expense of a healthy ecosystem has put us on this path, and I have little tolerance for that. The past few years have exposed many of these because of the deliberate acts of corporations bound to fossil fuels, and all who financially benefit from them.

I feel pretty certain that the Trump administration will be remembered for its blatant disregard for science, including climate change. The former Secretary of the Interior, Ryan Zinke, for one, was a thorn in my side for his two years of service. Not only was he inappropriate for the position, but he saw his role as one of exploiting versus protecting the very thing he was supposed to manage. Consider the policies pursued, enacted, or reversed during his term: opening public lands for mining and drilling, opening federal waters for offshore drilling, drastically reducing the bound-aries of national monument protected areas, reducing protection for endangered species, eliminating fines for killing migratory birds specifically protected by the Migratory Bird Act, reversing a ban on the importation of elephant and lion trophy items from Africa. And then there is the reversal of Obama's policy to reduce methane emissions from federal lands, and of course paving the way for drilling in the Arctic National Wildlife Refuge. In the end, he was forced out for financial mismanagement, not for crimes against nature or humanity.[9] Ryan Zinke is an avid fan of trophy hunting and darned if he isn't an endangered species. This chap-ter isn't long enough to allow for the environmental crimes and misdemeanors of Scott Pruitt and Andrew Wheeler, former heads of the Environmental Protection Agency (EPA). Make no mistake, none of these were honest mistakes of judgment.

We have lost our way as a country or culture if we somehow equate natural resources with negotiable currency. The bounty of planet Earth has always been its abundancy of resources, which isn't exactly shared equally. But this bounty is finite, with dimin-ishing returns as a result of climate change, something hardly imaginable to older generations. How are we to participate in a healthy economy without understanding the power of a purchase?

If we knew all along that the use of fossil fuel energy might starve us of clean air or contribute to a global climate crisis, would we have blindly purchased petroleum for our cars or coal, oil, and natural gas for heating and cooling our homes? When we aren't properly informed of our choices and their impacts, there is a bigger problem. That problem begins at the feet of large producers of these commodity goods. Imagine if companies were responsible for the full life cycle of their products including the hidden costs of downcycling and overall ecological impact. Would manufacturers make their products differently? We've barely tried. Here's a quote of mine from ten years ago.

We must stop consuming blindly and begin to pay heed to the process. There is an environmental price tag to every purchase. You just don't see it in the bill.

I don't want to be a participant in the plunder of our planet, but as consumers, we are, every day. We feel guilty throwing out products that can or likely won't be recycled. We don't like separating food waste from inorganic waste or PLA (polylactic acid) bioplastic from recyclable plastic.[10] We aren't inclined to buy products with a heavy environmental footprint or products with undecipherable labels. So, if you often feel guilty about consuming products, stop! I've been saying this for years. Don't feel personally guilty because that guilt is misplaced. We shouldn't condemn anyone for consuming because we are required to do so, but no one should go forward buying products without any sense of long-term impact. If you don't know where products come from, don't accept that. If you don't like how products are packaged, don't accept that. If you have no idea of the ecological impact of your purchased products, don't accept that. I've called countless companies over the years and learned a lot along the way. Make an informed purchase; that's all I can rightfully ask of you. And, don't feel guilty; make irresponsible producers feel guilty. Find out where these products come from and who is gaining from your purchase at the expense of the planet. Like my friend, Elaine Ireland, said, "Find out what's in the product and make conscious decisions about it. Just don't buy any ole thing out there simply because it's cheap. We're suffering the consequences of cheap on many levels. It's not working."[11] And I still say that our species is the only one that believes it can perpetually soil its nest—our planet—the only home we've ever known.

CHAPTER

Earth Day Reflection   2010

*T*oday, the 40th anniversary of the first Earth Day, arrives at an uncertain time for the environmental movement. The initial bipartisan effort, which was the brainchild of Senator Gaylord Nelson of Wisconsin, was a massive outpouring of support for what seemed so simple back then: preserving the planet from pollution and unnecessary harm as a result of unregulated industrial waste. Still, forty years later, it is worth noting that despite all of the public sentiment for cleaner water, air, and land for future generations, we have made baby steps. With the possible exception of the Montreal Protocol, which produced swift and thorough action assuring that our life-supporting ozone layer would not erode, we have had few global environmental victories. What many Americans quickly grasped in 1970 has become far more confusing today. Why are we still losing the carbon battle? Why *do we still have days of unhealthy smog levels? Why are we still fouling our oceans and waterways? Why are we not preserving our fellow species? Most importantly, why aren't we all involved in grassroots efforts to bring greater awareness to the multitude of issues facing our planet Earth?*

I think it is time to renew our goodwill toward the environment. Get reacquainted with why we first embraced the concept of sustainability. Let's not leave this movement in the hands of compromised corporate leaders or risk-averse politicians. Without strong public action, like that of the original foot soldiers of the '70s, we cannot expect to see radical change. Just think if the new decade created the next Environmental Protection Agency, Clean Air Act, Endangered Species Act, or Earth Day. We are of one political stripe when it comes to the environment. We are of one mind when it comes to what we can achieve for our descendants. Let's spend this Earth Day renewing our memberships to bluer oceans, greener grasses, clearer skies and our resolve to sowing seeds one young mind at a time.

Senator Gaylord Nelson and the Origin of Earth Day

The Earth Day teach-ins, a concept of Senator Gaylord Nelson of Wisconsin, went viral in a bipartisan effort to promote environmental conservation in 1970. Nelson was a passionate environmentalist who had been lobbying for a range of ecological measures in Congress. His reputation to that effect was well-known in Washington by 1969 when he developed the idea of an environmental teach-in. In fact he didn't even refer to the event initially as Earth Day.[1] Nonetheless, Earth Day will always be his. The beauty of Earth Day One is the nonviolent, mostly bipartisan message that was shared intergenerationally and in the spirit of hope. A movement spread by a willful army of young and old, exhausted by war, racial inequality, cultural and sexual restrictions, and environmental disregard, that had no political battleground. The organizers, including Denis Hayes, noted environmental advocate and educator, succeeded in gaining participation in the millions. Venues around the country held events with speakers as diverse as Gaylord Nelson and Allen Ginsburg, Barry Goldwater and Barry Commoner, Edmund Muskie and NYC Mayor John Lindsay with actors, Paul Newman and Ali McGraw, in tow.[2,3]

It did receive its fair share of criticism for its naivete and for ignoring the chasm created by the Vietnam War movement. I think that's unfair to its founder, Senator Nelson. He had been a vocal opponent of the Vietnam War and he was considered one of the most progressive senators in the country. He was also quite effective in gaining bipartisan support for issues of great importance to him, and he was well regarded on both sides of the aisle. His genius was in making Earth Day a nonpolitical concept that could rally mass support.

The tepid overall support from the African American community was more a consequence of the concurrent battle for civil rights than a rejection of Earth Day at the time. Nowadays, certainly, we know that we cannot separate civil rights from environmental rights. They are interdependent since persons of color, in many cases, are the front lines of climate change and the least responsible for it. The significance of Earth Day, 1970, was perhaps best grasped by one of the twentieth century's finest journalists: Walter Cronkite, anchor of *CBS Evening News* for twenty years. His final remarks on Earth Day were clear warnings that we may fail to take this seriously but only at our own peril, which was a clarion call to act before the problem becomes too big to overcome. In less than five minutes he prescribed the future. Maybe he knew us too well. Walter Cronkite was the voice many Americans chose to trust, and he summarized the importance of protecting the planet in such earnest terms. I recently viewed the replay of his concluding segment on the first Earth Day—in retrospect, it is disturbing all these years later.[4]

A Reassessment of 4/22/1970

To be fair, there were many scientists already engaged with the government to address energy policies and climate by the time of Earth Day, 1970. They saw Earth Day as more of a public rally than a serious legislative challenge to governmental policy on the environment. One of them, Dr. George M. Woodwell, founder of Woods Hole Research Center (now known as the Woodwell Climate Research Center), had been involved in environmental policy issues well before April 1970. George is one of Earth's heroes, and it is a privilege to still be in contact with him. His commitment to environmental rights runs long and deep.

I was fortunate to meet Dr. Woodwell in 2002 as a result of my green furnishings company, Olive Designs, and subsequently, in 2014, to interview him for my first book. I queried him in 2019 about his thoughts of the events of April 22, 1970, and he replied, "In 1970 I was thoroughly engaged in public interest law having been involved in founding both EDF [Environmental Defense Fund] in 1965-1966 and the NRDC [Natural Resources Defense Council] in 1970. We were busy forcing the government to obey its own laws. 'Earth Day' seemed late and frivolous by that time, no matter that it was a great idea and brought much more interest to environmental affairs."

I get that. And I would concur with Dr. Woodwell in that if we don't shape policy by our actions, then we aren't making any substantive improvements. He, like Walter Cronkite, encapsulate why the first Earth Day was so special and yet so underwhelming in long-term progress.

None of the original participants would be pleased to know that we have not effectually changed the course of events. Climate chaos, as we refer to it now, is upon us. Fifty years of Earth Days haven't produced the results the founders envisioned. And George is right that it was a great idea, but neither he nor Senator Nelson could have known that in 1970. But knowing of Dr. Woodwell's concerted efforts and commitment, I can confidently state that we haven't achieved anything close to what those early activists sought. If we had listened to the scientists of the time versus following the 1980s-led, anti-environment policies, we may not be so imperiled. For all of the Nixon and Carter Administration flaws, we might have paid more heed to them when it came to renewable energy and conservation.

Gaylord Nelson really had no idea then how big the 22nd of April, 1970, would grow to be. Considering the media attention that it did get, it makes perfect sense. It was ever-present, showing up on countless billboards and banners. The *New York Times* even published an advertisement of Earth Day that stated its purpose:

> Earth Day is a commitment to make life better, not just bigger and faster; to provide real rather than rhetorical solutions. It is a day to re-examine the ethic of individual progress at mankind's expense. It is a day to challenge the corporate and governmental leaders who promise change, but who shortchange the necessary programs. It is a day for looking beyond tomorrow. April 22 seeks a future worth living. April 22 seeks a future.[5]

In review of the first Earth Day, there were 20 million followers who participated in some form or shape. From kindergarteners to senior citizens, people marched, danced, drew, listened, spoke, sang, and honored the health of our planet in a myriad of ways. It was not limited to the United States, either. We, today's environmental protesters, aren't as naïve nor as patient but we sure have a lot more at stake now.

The Last Wrong Dictionary

In 1972, when I was nine, I wrote an *A-to-Z* booklet on the environment to honor the second anniversary of Earth Day. Finding the booklet again recently also reminded me why Mom made me practice cursive day and night. Here are excerpts from that booklet entitled, *The Last Wrong Dictionary* [an ironic title]:

B is for the blood of a dead fish.

C is for care that we don't have.

F is for food that is dying off.

H is for help, which is needed desperately.

M is for man who is the only one who can help.

O is for ocean that is now polluted.

U is for US our generation.

V is for vessels filled with trash.

Y is for the young generation that is going to help.

Z is for Zamerica [the alternative].

B is for the blood of a dead fish. We sure saw our share of dying fish in 2018. The red tide that struck Florida's Gulf Coast brought dead marine life to the shores en masse. The severity of this cycle of red tides has many culprits including higher ocean temperatures as well as the leaching of chemicals into bodies of water. Once again, the consequence of pollution isn't factored into the costs of its usage or production. Pesticides, herbicides, synthetic fertilizers, agricultural runoff and warming ocean temperatures are a deadly combination. Whether they are to blame for the year-long red tide of 2018 along Florida's Gulf Coast or not, most marine scientists will tell you that they play a part. Nutrient-rich waters are a breeding ground for algal blooms.[6] Irresponsible coastal development can encroach on the very swamps and mangroves necessary to filter these chemicals before they enter a body of water.

Fish kills aren't a new phenomenon but scooping them up by the thousands of tons isn't common nor are red tides that last for a year or more. Scientists now conclude that the ocean's changing temps and pH balance equates to more acidic waters that are killing marine life, including coral and shellfish.[7] And despite the fact that all of the world's fisheries are in decline, the beaching of dolphins, whales, and other marine species suggests a more sinister outcome.

It could easily be argued that we have not shown enough care for the one planet that supports life as we know it. C is for taking care to prevent the worst consequences of environmental chaos that we are beginning to see today on a worldwide scale. The scenes of desperation by the least privileged of us is about as clear as it can be that we haven't shown enough care for our earth and its inhabitants. Access to safe drinking water, for example, should be a universal right, yet one child dies every twenty-one seconds as a result of unsafe water.[8]

We're not just talking about poorer nations around the globe either. Right here in America we got to see the brown water of Flint, Michigan, and that of Newark, New Jersey. The state of Michigan was even willing to put the health of their citizens at risk. Anna Clark, author of *The Poisoned City: Flint's Water and the American Tragedy*, summarized the water crisis of Flint:

> The chronic underfunding of American cities imperils the health of citizens. It also stunts their ability to become full participants in a democratic society, and it shatters their trust in the public realm. Communities that are poor and communities of color—and especially those that are both—are hurt worst of all.[9]

Flint native, documentarian Michael Moore, echoes this same sentiment, and as a passionate mouthpiece for the citizens of Flint, has brought much-needed awareness to the crisis. He is also rightly identifying that lead ingestion at any level is going to cause health issues for years to come.[10] In the example of Flint, C could also represent cover-up. Care has always been a fundamental component of sustainability, one that should be universal in light of the overwhelming loss that we'll be forced to endure.

I often wonder whether we truly don't care about a healthy, diverse habitat or whether it has to do with forethought. Care is a factor of forethought. We humans have redefined wildlife in a matter of a few centuries and not taken pause long enough to halt this course of action. That is a demonstrative lack of forethought. It isn't as if a group of famed biologists and naturalists weren't warning us for the better part of a century. Were we really listening or is it uglier than that? Do we really care? Does it matter to a trader on Wall Street whether Sumatran rhinos go extinct thousands of miles across the globe? How does a golden toad dying off in Costa Rica affect a wheat farmer in Kansas?

But losing species isn't unrelated to planetary decay. We are poised to witness a massive loss of species as habitats shrink and irreversible climate change becomes our reality. The everchanging landscape of the future isn't going to be so hospitable to many species, which is another way of saying that the loss of habitat is the mechanism for mass extinction. These extinctions, which may spare us, are a complete failure in concept and design. For an evolved mammal, capable of landing objects on Mars, eradicating diseases, and communicating digitally, the idea that we can't protect the integrity and diversity of the only planet hosting life in our galaxy shows a complete lack of forethought. As such, it also lacks care. That we have drained our planet of its living resources and not paused to change our course of action seems incongruous with a caring and compassionate society.

C unfortunately for us in 2020 also represents COVID-19 or officially, SARS-CoV-2. But amidst the horrors of losing close to a quarter of a million Americans, and over one million people globally, by mid-November 2020, we saw incredible acts of dedication and compassion. To have witnessed this tragedy is to never forget that there are beautiful people everywhere who are willing to sacrifice their well-being to care for others. We've seen doctors, nurses, first responders, and essential workers die in the line of duty. That is more than any of them signed up for and more than we should ask of them. Doctors and nurses became the de facto families everywhere of men and women breathing their last gasps of air, unable to bid goodbye to their loved ones. Notwithstanding the fact that it didn't have to come to this level of suffering, people demonstrated profound acts of kindness. Hearing their stories of heroism should mete out a better response from the rest of us.

Caring for one's self cannot be disparate from caring for others. Successful societies rely on coexistence and compatibility. It is time we get back to that. That is, in concept, the binding tie between living today and life in perpetuity. We can't be disengaged from caring for the environment lest we're willing to jeopardize our future. Care has never left us, as we have seen in 2020, but we must extend it to that which sustains us and to that which we may never know. Just like doctors and nurses did for the countless dying strangers of 2020.

F means food, but I am now convinced that there is still enough food to feed the planet. Having struck up a friendship with Frances Moore Lappé, author of eighteen books including the bestseller *Diet for a Small Planet* in 1971, she has shaped my opinion on feeding the world.[11] Frankie makes a compelling argument that what is lacking is democracy, not food. When food is not distributed universally, and when land is not efficiently utilized, scarcities are more a factor of greed or misappropriation. Furthermore, appropriating a third of all arable land to grow grain for feeding cows has taken on a new problem: greenhouse gases. The Intergovernmental Panel on Climate Change added another challenge to status quo by indicating that reforestation is critical to reducing CO_2 emissions, which is compounded by the clearing of land and the subsequent methane production from the meat industry.[12] Converting diets to plant-based protein from animal protein is a big step to combating climate change.[13] Don't take it from me but from Pat Brown, founder of Impossible Foods, who says it best, "The use of animals in food production is by far the most destructive technology on earth."[14] Plant-based burgers are in full swing, and I've sampled them all. Burger King has made a big step by introducing the Impossible Whopper, a project only a Californian could conceive. I've eaten plenty of them, and for someone who hasn't eaten beef for almost forty years, I don't see what the big deal is. I can't tell the difference between a beef burger and an Impossible Whopper, other than the 3x price tag, which is the key hurdle Brown recognizes his company must overcome.

It is worth noting that Frankie sold over three million copies of *Diet for a Small Planet*, and my Earth Day booklet written a year later failed to catch on. Mom and Dad didn't throw it out, so that's something.

H is for help, which is needed desperately. There's a good chance I needed help in spelling *desperately* in fourth grade. It is interesting to consider where we would be if we had the help in 1970 to take the threat of climate change seriously. It isn't that we didn't know back then that the combustion of fossil fuels had the potential to alter the planet's temperature. This was laid out in the 1960s if not before. The fossil fuel companies knew it too. Some even discussed it internally while ramping up coal, oil, and gas production.[15] What makes this so disturbing is that the perpetrators were seated at the table. Most major fossil fuel companies were meeting with politicians and environmental groups to discuss alternative energies and the potential for climate change thirty or more years ago. Their feigned interest is hard to forgive today. Imagine that your commitment to science and education makes you an expert on the very thing your organization chooses to hide from the public. What good did a first-rate education do for you if you are fundamentally dishonest?

I remember visiting Borneo on the last leg of a sustainability consulting stint in Indonesia in 2008 and meeting an American at the Balikpapan Airport. The purpose of his trip: oil exploration. All I could think of at the time was that Earth has no chance of fending off fossil fuel companies and that there is no end to corporate greed. Help and hope go hand in hand though, I've since realized.

The global efforts of environmentalists and concerned citizens has mushroomed since 2008. Climate change is a threat regardless of age and profession, but I am overwhelmed by the engagement and activism of teens and twenty-somethings. My sincere thanks go out to educators everywhere who have spent time taking on the subject of climate change. We have a new demographic group unwilling to be fed any more lies and excuses. They're out there at city halls, conventions, political offices, banks, businesses, airports, the UN, grocery stores, college campuses, and just about everywhere else in between. They chant, cajole, and engage, in numbers I've not witnessed before. They impact policy; they influence political races; and I would argue they have been the arbiter of the Green New Deal. To suggest that a Green New Deal is impractical to a generation who bears the brunt of runaway climate change is shortsighted and short of compassion. What economic indicators factor in a force as potent as habitat destruction? The billions lost due to climate-related incidents so outweigh anything short of global nuclear war. To see political will bending in the direction of a green youth movement is a hopeful sight. I see help on its way but brace yourself if you think the fight is going to be easy.

M is for man, who is the only one who can help. I am pretty sure that as a nine-year-old I didn't comprehend gender discrimination. I used *man* to indicate humankind—I just wanted to get that out of the way. It is a conundrum though to know that we are the culprit and the solution. We found a way to alter the planet's chemistry and disrupt its patterns in a blip of time. This is a statement I have used for years:

We are currently conducting a unique experiment on the only known habitat that can support life for our species.

Knowing that, what stakes could be higher? This leads me to the ugly part of this experiment. We humans haven't treated this problem with the seriousness it requires. Having given a lecture or two on the modern environment movement, the problem has grown as we have kicked the can down the road since the 1970s. Richard Nixon was many things but an ardent environmentalist he was not. Yet he gets undue credit for advancing legislation that has proven to be critical to today: the formation of the Environmental Protection Agency, The Endangered Species Act of 1973, Safe Drinking Water Act, Council on Environmental Quality, creation of Occupational Safety and Health Administration (OSHA), and other landmark achievements. I am conjuring thoughts of Senator Gaylord Nelson again. The timeline makes a fairly sharp drop in successful legislation in the 1980s.

President Reagan for all his California connections and Western wear was far less interested in climate change than the peanut farmer who preceded him. Jimmy Carter appointed a top-notch group of scientists in 1979 to provide him with detailed analysis of energy usage and environmental impact. One of those scientists, incidentally, was Dr. Woodwell.[16] This is the point, in my opinion, where the train got derailed. A comprehensive study done for Carter's Council on Environmental Quality made it abundantly clear that we would see planetary conditions change if we did not curb our fossil fuel energy usage.

There was a time when climate politics wasn't the domain of one political party or the other, by the way. The late President George H.W. Bush even campaigned on the issue and was an actual if a bit lukewarm supporter until he realized that it didn't fundamentally move his political base. And although his campaign promise of using the White House effect to combat the greenhouse effect didn't amount to a key environmental policy in his administration, he did bolster efforts to strengthen the Clean Air Act.

One of the better books I've read on the topic, *Losing Earth: A Recent History* by Nathaniel Rich, outlines this political paralysis of the past forty years.[17] The blame for the current environment predicament is broad enough to spread around. But only we humans could make preserving our planet an economic argument versus an existential one.

One of my favorite cartoonists is Pat Hardin, another Flint, Michigan native, whose work has appeared in many publications including *Saturday Evening Post*. One of his cartoons that I particularly like is an evolution timeline with captions starting with a fish making its way to land to a four-legged critter to a slightly upright creature resembling an ape all with the same caption: "Eat, sleep, reproduce." The final figure in the timeline is modern man with an updated caption: "What's it all about?"[18] I guess we forgot the basic foundation of existence somewhere in the late Holocene epoch. It certainly makes sense to me in the face of climate crisis.

V is for vessels filled with trash is very real; up until January 2018, China was processing our recycled waste. After more than twenty years of accepting our and other nations' waste, they decided to end this practice at the close of 2017. These container ships are now grounded and our domestic trash is going nowhere. *V* for vessels filled with trash are coming home to roost as I write this. No nation has stepped up to take our refuse, although I am concerned that Indonesia may be accepting a portion of it.[19] One disturbing incident I recall from spending time in Indonesia was the sight of rejected Mattel toys from China. Traces of lead paint were found on Mattel toys made in China and rejected by the U.S. market back in 2007.[20] No nations should be exporting trash from here on out. Seal your environmental borders by taking care of your products; no, not when they lose their use but when they are designed. We must find the curiosity to change our methods of production and, simultaneously, our waste. *V* in 2020 also stands for vote. And if you cannot determine a candidate or an issue that motivates you to vote, it matters to other species. Larkspurs aren't going to bloom in voting booths but their protection might.

Finally, I wonder what Senator Nelson would think today? He lived to the ripe old age of 89, ultimately earning the highest honor a U.S. civilian can receive—the Presidential Medal of Honor—in 1995. If we had more senators like Gaylord Nelson, we might not be in our current situation. And concepts like the Green New Deal wouldn't strike us as desperate or obligatory. He lived long enough to witness political cowardice and the corporate hijacking of governmental policy. He would be discouraged by our lack of international leadership on climate change. From Kyoto to the Paris Accord, U.S. leadership on the topic of climate change has been conspicuously weak, nonexistent as of the Trump administration. Rest in peace, Senator Nelson, you fought bravely.

Y is for the young generation that is going to help. That sentence doesn't quite have the same meaning today because it is our youth who are the drivers of the current modern environment movement. Their voices are louder than the rest of us, and with good reason. On Friday, March 18, 2019, over a million students from 100 nations or more skipped school in a school strike for climate. They are taking on the problem with greater urgency, and we should actively help them lest we look more complicit than ever. It will be exceedingly difficult for climate deniers to attack sixteen-year-olds on these issues. They haven't had decades to sit idle while climate change and its effects have intensified.

Sweden's Greta Thunberg has become as recognizable as Hollywood's A-list actors. On December 4, 2018, Greta Thunberg calmly and courageously addressed the world leaders at COP 24, the 24th Conference of the Parties of the Climate Change Convention, where she said, "The year 2078, I will celebrate my 75th birthday. If I have children maybe they will spend that day with me. Maybe they will ask me about you. Maybe they will ask why you didn't do anything while there still was time to act."[21] I was younger than Greta when I wrote the *Last Wrong Dictionary*, and I was asking the same question back then.

Today's young eco-warriors are a far more active and unified group, and social media offers them a massive mouthpiece that we lacked in 1970. They understand the political power they can wield. Believe me, they are ready to fight for their future. There are a number of potent youth organizations pushing for action.

Greta has rightfully noted that climate change also represents a matter of fairness and equity. When one small body gains at the sake of a larger body, it is imbalanced and unequitable. Fossil fuel producers have prospered without the economic consequences of the pollution to the planet that they have caused and the climate chaos they have helped foster, at least until recently. Poor nations lack funds to adequately prepare for coastal erosion, agricultural loss, fishery collapse, flooding, freshwater shortages, species loss, a reduction in tourism, trade imbalances, pestilence and disease, and lastly, the instability of forced migration.

And although we Americans share some of these grave consequences of climate change, we are still a rich nation with a wealth of resources. This is not a matter of resource scarcity here in America, it is purely an economic one. Until we level the playing field between renewable energy sources and fossil fuel-based ones, we'll continue to drag the problem out. It was that way in 1970 and it is the same today. No one serious about mitigating climate change wants to continue to subsidize the extraction of oil. What economic genius determined that a wise policy would be to charge the people for discounts on the very product contributing to Earth's decline? And this subsidy doesn't even factor in the many unintended consequences of fossil fuel extraction or hidden costs including the long-term impacts of environmental degradation.

Z is for Zamerica. Of course, Zamerica isn't a real place, and it's a place we don't want to create. It's likely that at nine I couldn't think of any other word starting with Z but based on the other responses, it did connote a failure at some level. I was a young boy with unusual concern about our environment, albeit from a juvenile's perspective. I remain concerned, maybe alarmed, since much of what has happened these past fifty years to our planet isn't good. Far too much. Maybe we are living in a world closer to Zamerica than the America of 1972 but let's not get to Zamerica.

3

CHAPTER

A Vote Is Heard in the Night 2011

*T*his anniversary of the first Earth Day is marked by an act of indignity toward the spirit of the movement and to those intrepid leaders who rallied millions in an attempt to influence politicians and a government that they mostly distrusted. Forty-one years later, we have reason to distrust them again. One of the critical building blocks of a healthy natural environment is species diversity. Those early environmental activists recognized this and helped establish the Endangered Species Act of 1973. Imagine if you can, every time you pulled out paper currency from your purse or wallet, you saw a picture of our recently extinct national symbol. Or imagine that a bison, a denizen of our former Western landscape, was extinct in one of the great mass slaughters of the Cenozoic era. And to think, these are species we actually revered.

In a historic, if not inconceivable, sleight of hand, the government has delisted the gray wolf from federal endangered species protection in Montana and Idaho. Since the gray wolf also resides in Oregon, Wyoming, and Washington, I hope someone fully explained the concept of state borders to them since they are not known as homebodies. This act, brought forward by Senator Tester of Montana and Congressman Simpson of Idaho, was attached to the recent budget bill (HR 1473 of 2011) signed last minute to avoid a government shutdown. How this senseless act is going to significantly help balance our budget has yet to be disclosed.

The idea of delisting a species just reintroduced to the region in 1995 is a cruel twist of fate for those seeking to restore the region's natural ecosystem. By all accounts the gray wolves' return has had a positive effect on the region including the repopulation of important native hardwood species like aspens.

Of course, this wouldn't be the first time the area rid itself of wolves. Citizens of the region were successful in eradicating wolves early in the twentieth century. Ostensibly, removing federal protection doesn't assure extinction for the species but turning over control of the species to states politically dominated by ranchers and hunters doesn't bode well for them. Soon, these very hunters and ranchers will be targeting wolves, make no mistake about it. With declining breeding pairs in Yellowstone National Park, where they are still federally protected, one can imagine what will happen in states where they aren't. Without safe buffer zones, culling of the species could have a permanent impact.

And don't be fooled by reports that wolves are running rampant, devouring livestock at every corner. According to a 2006 report by The National Agricultural Statistics Service, more livestock were lost to domestic dogs and vultures than wolves. Even cattle rustling, yes, cattle rustling, was responsible for five times more cows lost overall versus wolf predation.[1] So how did *we take this giant leap backward? It is another act of politics manipulating science. In a period where scientists are warning of a pending mass extinction of species, delisting gray wolves is an ecological paradox.*

It is time for us to march again, to rally again. It is time for us to promote science over politics, to speak out to our elected officials on behalf of those who cannot. To carry the baton of those brave Americans who enacted the Endangered Species Act (ESA) into law. On this Earth Day anniversary, few should be howling victory for the delisting budget maneuver pulled by Montana and Idaho politicians. We can hold out faith, however, that they will encourage their constituents to build eco-friendly straw bale houses now that the big bad wolf won't be huffing and puffing anymore.

Top Predator or Pariah?

One of the most misunderstood species in the United States is the *Canis lupus*, the gray wolf. From childhood we were taught to be afraid of the big bad wolf. I've never seen a wolf in the wild, and I bet most of you haven't either. They are supremely gifted hunters and survivors, a fact that seems to add to their lore. The big bad wolf wasn't pursued so vehemently because he ate Grandma or tried to do the same to Little Red Riding Hood, he was nearly eradicated because he was a perceived threat to ranchers, farmers, and hunters. Naturalists, foresters, and ecologists all recognize how important a predator like the wolf is to maintain a precious balance between herbivores, omnivores, and carnivores. Wolves were hunted close to extinction in the nineteenth and twentieth centuries because of a fear derived more from children's books versus reality—a fact I pointed out in my Earth Day essay for 2011.

According to Shawn Cantrell, Vice President of Field Conservation at Defenders of Wildlife, an expert on wolf management, "No other critter has generated as much visceral hatred as wolves."[2] Defenders of Wildlife is one of the principal nonprofit organizations responsible for wildlife and habitat conservation in the United States, and they have, as much as any other organization, taken on the case of gray wolves. Their work includes education and legal defense of species threatened by habitat and population loss since 1947. As it relates to wolves, that has kept them busy. Shawn made it pretty clear to me also that U.S. Fish and Wildlife Services has no interest in the problem of managing wolves, abdicating the responsibility back to the states. I can corroborate that since my initial conversation and subsequent request for information with a representative from U.S. Fish and Wildlife Services yielded no response. And borders do matter whether wolves recognize them or not since den locations determine which state has responsibility for a pack. So, although Idaho authorizes wolf hunting games for the entire family, the states of Washington and Oregon have worked to protect their numbers.[3,4] Moreover, California has recently taken the step to uphold an earlier legal ruling to protect the state endangered species laws since wolves have begun to cross back into their state as of 2014.[5] In contrast, Wyoming has often had their policies legally rejected in the past due to inadequate state management measures.

Having spent his career protecting endangered species, Shawn believes that if given half a chance, wolves will make it, and he remains cautiously optimistic. Part of his optimism stems from how certain states manage the wolf population; his concern revolves around the delisting of a species just twenty-four years after reintroduction. A healthy stable population can take a century to be fully established, so the decision to start culling the wolf population is risky. It would be beyond cruel to repeat our wanton slaughter of *Canis lupus*, but given the opportunity, ranchers will adopt the policy of "shoot, shovel, and shut up," which is akin to poaching, Shawn explained.[6] Our passionate sentiments have certainly clouded judgment when it comes to wolves. Childhood fairy tales be damned.

I have plenty of friends who are hunters, and although I abhor the idea of shooting a living, breathing creature that poses no threat to me, I don't have any animosity at all toward them. Nor would I expect my buddies to approve of all my activities but that doesn't and shouldn't impact our friendship. I am going to question, however, the family bonding experience of treeing a mountain lion, a cruel hunting technique, since I mean, really, the dogs did all the work but I guess they are part of the family.

We have a deer problem, a tick problem, a tree problem, a road hazard problem, but more importantly a perception problem as a result of eliminating wolves. Deer and elk populations in many states rose rapidly once wolves were reduced in numbers. We couldn't imagine that our misplaced fears could cause an ecological imbalance but that is exactly what happened in many regions of the country. Culling ungulates like deer and elk by relying on hunters throughout the United States is not natural selection. Wolves keep the population in check by catching the most vulnerable of the species. The overpopulation of these hooved creatures means that they have changed the ecosystem in many places. Anytime that happens, consequences occur. The precious landscape of Yellowstone National Park changed, it was determined, when wolves disappeared.[7] In so many instances, when you remove a top predator from a habitat, the ecological balance gets tipped. The elk and deer in Yellowstone lacked the fear of predators and began to graze in the open and consume seedlings, which changed the prospects of aspen and willow trees, for example. And that set off a chain effect of ecological anomalies for the area.

The reintroduction of wolves into many of their native habitats was a fight for the ages. Bitter fights between ranchers and environmentalists went on for a decade or more. In 1995, legislation was approved that would compensate ranchers and farmers for losses from wolf predation after their reintroduction into Yellowstone. In most cases, the specific state compensates the claims of livestock loss, but each state has its own policy. The government will also pay a rancher or farmer's claim in certain cases. It is worth noting that the number of claims remain quite small in perspective to the size of livestock herds. Wolf packs were never big enough to have warranted their expulsion in the first place. In nature if Red Riding Hood wandered into the forest to pick flowers, a healthy wolf would pick up her scent and scatter. Wolves, like their cousin, canines, have a sense of smell superior to ours—superior by many thousands of times. By the time they smell us, they're nowhere in sight. So why did we choose to extinguish them? Fear. A pack of wolves is a slick, efficient, fast-paced hunting machine. They may be one of the most efficient hunters that ever swam, crawled, soared, or scampered on Earth. Even more reason to respect them not subtract them.

Elemental Fear

A rancher's son
learns to hunt at eight,
but prejudice and precedent
need no tutor.

A stray from the pack
peels off a meal
from the pavement,
his appearance owed
to reintroduction.

Tufts of gray fur
are swept away with
the Wyoming breeze.
No one chalked the murder site,
no one suffered the truth.

Is best to keep the
doors locked
at night anyhow,
those weren't wolf howls.
(2004)

Endangered Species
and Land Management

The combative relationship between ranchers, hunters, and farmers, and the ESA reared its ugly head as a result of the reintroduction of wolves into Yellowstone National Park. The reaction to the reintroduction of American bison or buffalo into Yellowstone didn't spark anything similar to the reaction to gray wolves. Of course once a bison sets foot outside of Yellowstone Park, they can also be shot indiscriminately. Yet their population is stable and Yellowstone has the biggest herd of wild plains bison on public lands in North America. It is another sign that we can rehabilitate species and reintroduce them into native grounds without disruption. Wolves do come into contact with ranchers, farmers, and landowners in the areas surrounding Yellowstone National Park. These folks fear for their livestock as well as for their pets when outdoors. There is no doubt that a wolf pack could kill a healthy cow, sheep, dog, or family pet. These occurrences are rare, as I noted previously, and hardly worth the eradication of wolves but the battle has raged on regardless for decades.

Wolves were reintroduced during the Clinton Administration and on the back of Secretary of the Interior Bruce Babbitt. Babbitt, a descendant of the environmental modus of Gaylord Nelson, saw to it that gray wolves were brought back to their ancestral homelands. It immediately made him an enemy to many ranchers, farmers, and hunters throughout the affected area. They were for the most part opposed to keeping gray wolves listed as a threatened or endangered species since delisting them from federal protection meant that states would be responsible for managing stable wolf numbers, which in some cases is akin to the fox watching the henhouse. Wyoming at one point proposed the killing of wolves from one end of the state to the other. Now that doesn't sound so sporting, does it?

The George W. Bush administration was particularly interested in delisting gray wolves in the Northern Rockies. From 2003 to 2009, the U.S. Fish and Wildlife Services of the Department of the Interior continually proposed eliminating any federal protection, allowing the indiscriminate killing of the Northern Rockies' wolves including those that wandered outside of Yellowstone National Park. We can only hope that Wyoming hunters were more like Dick Cheney, the pathfinding scout and expert marksman from Wyoming and Bush's V.P., than, let's say, Daniel Boone. Cheney had a better chance of racking up another hunter than tracking and killing a healthy male gray wolf. One can see his effect on his daughter, Liz, representing his old congressional seat in Wyoming. She made it her priority to prod Secretary of the Interior, Ryan Zinke, to delist wolves from Wyoming and planet Earth in 2017. Zinke was all too pleased to oblige. The Trump Administration finally prevailed on the big bad wolves and removed gray wolves from the Endangered Species Act in October 2020. Defenders of Wildlife along with five other conservation organizations went forward and sued the Trump administration on January 14, 2021 to overturn this ill-conceived ruling.[8]

Although the wolf population has remained stable in certain regions of their original habitat, their future is anyone's guess. The jury is still out whether the wolf will survive. My sense is that we can manage their broad roaming existence in Yellowstone National Park, and we can bring them back into other less-populated areas.

By the way, if wolves exist in an area, the coyote population there is also affected. It is estimated that predators like wolves reduce coyote numbers by half.[9] Coyotes have been urbanized, which also separates them from wolf packs. Wolves are not really killing your cats and doglets (dogs in small packages); coyotes are, though, for sure.

The ESA, put into legislation in 1973, was an expanded version of a similar law established in 1966. Its general purpose is to prevent and preserve endangered and threatened species and associated habitat, or as the U.S. Fish and Wildlife Service states it, "The purposes of this Act are to provide a means whereby the ecosystems upon which endangered species and threatened species depend may be conserved, to provide a program for the conservation of such endangered species and threatened species."[10] The modern version of the ESA also includes the Convention on International Trade in Endangered Species (CITES), which bans the import and export of endangered species. You know those things that are a must to have like rhino horns, elephant ivory, hawksbill turtle shells, and bear paws. What home doesn't need these, I mean, really? Even though America is guilty of playing in the black market for real, Asia is where the problem is so troubling. Because of antiquated but sacred customs, many nations trade in endangered species for medicinal reasons. Ridiculous reasons as we will see in chapter 4. Do the Chinese really think that a fur seal's penis will make them more sexually potent?[11] Who comes up with this stuff, anyhow? If you think you need a fur seal penis to procreate then I don't want you to procreate at all.

The ESA became controversial primarily because of land management rights. Because habitats of an endangered species may be wider and larger than anticipated, humans can be impacted. Why should a business cease development over a damned owl? Yes, that spotted owl. The controversial decision to protect old growth forests and establish a protected habitat for northern spotted owls became national news in the 1980s and 1990s. Environmentalists and the timber industry squared off over land management rights. The loggers decried the end of their livelihood while conservationists lobbied for the rights of a declining species. The episode boiled over and the Clinton Administration got involved and a policy called the Northwest Forest Plan was enacted in 1994.[12] Although logging of old-growth forests has been curbed, the spotted owl has not recovered. Historically speaking, we humans just don't share resources all that well. And in this case, neither do barred owls. Their recent presence has also affected the survival of northern spotted owls because they are a far more prolific species. One can make their own conclusions but I'll share mine: logging old-growth forests in public lands is not a long-term solution for traditional industries like timber. I fervidly believe that there is a long-term place for the timber industry and logging in general. There will always be an industry for the cultivation of lumber. Wood is an essential raw material for many industries, including the building industry and furniture, a part of my livelihood. But let's keep the old-growth forests intact and preserve these vibrant ecosystems. The more biodiverse a habitat is, the healthier it is. Land management can include the planting and harvesting of sustainable species of wood. Sustainable forestry management programs, including Forest Stewardship Council and Sustainable Forestry Initiative, have proven to be successful in many regions of the globe.[13] The United States has increased its forest cover for many years, which is a positive trend since trees are such a key component to mitigating climate change. Keeping public lands intact increases the activity of natural carbon sinks, and in light of the battle to reduce CO_2 emissions, that's kind of important.

Will the Wolf Survive?

If the wolf can make it, it bodes well for all species. If we fear reintroducing key predators like wolves it may lead to unintended consequences. We must deny our unfounded fears and respect the well-being of species we may never understand like wolves, snakes, sharks, and pangolins. I don't fear pangolins—unless they are the intermediary species for COVID-19 transmission—but, man, how could a scaly, nocturnal creature that eats only insects become the single most illegally traded species on the planet? Beauty is in the numbers. Let's exalt wolves, not fear them; if we do, then we will hear wolf howls in the wild where they belong and not in our nightmares. I'll leave the last word on the subject to my friend, singer and songwriter and artist extraordinaire, Louie Pérez, of the band, Los Lobos: "Running scared, now forced to hide. In a land where he once stood with pride. But he'll find his way by the morning light."[14] Let's hope we find ours too.

CHAPTER 4

What Must They Be Thinking? 2012

*I*t is the naivete of us nonbiologists to ascribe human feelings to our fellow species, a phenomenon called anthropomorphism. I do it often because in my childlike wonder of nature I believe that an adult gorilla conjugates emotions in greater depths than we know from scientific studies. Maybe I humanized the protagonist from the powerful novel, Ishmael, by Daniel Quinn, more than was intended by the author.[1] But other than the conservationists dedicated to these critically endangered species of the world, who else is going to look out for them and who else is going to tell *their stories? We humans perched at the top of the evolutionary pyramid have made it increasingly difficult for many signature species to coexist. I decided, this forty-second anniversary of Earth Day, to offer my implausible interpretation of what three signature species are thinking in today's precarious times.*

The various species of elephants are universally well known as one of the largest mammalian creatures as well as one of the most intelligent in existence. Recently it has been observed that elephants cry at the death of a family member, and the herd will join together in a ritualistic formation as an expression of loss and respect for the deceased. If they can perceive loss in such profound ways, it isn't hard to theorize that they are well aware that they are now prisoners of a planet they once roamed freely. Intelligent creatures don't accept enslavement easily, a lesson not lost on Homo sapiens. They must know that sustainability doesn't apply to them. I don't want to see elephants resigned to zoos and circuses, and I'm pretty sure they don't either. So, the next time you see elephants in public lined together, you might wonder if it is a mass funeral.

The state of our oceans is increasingly impaired, and the large cetaceans must know it too. Our crude understanding of the songs of humpback whales gives us very few clues. Humpback whales sing complex songs, albeit among males only, and communicate effectively in high and low frequencies. Perhaps these songs are nothing other than whale pickup lines but I doubt it. In our globalized world of commerce, we have introduced a new invasive species called container ships. With the amount of cargo ships outnumbering remaining humpback populations threefold, these songs must be getting a little desperate. With oceanic pollution and warming temperatures on the rise, the songs of the humpback are the new dirges of the deep: planetary distress signals.

By now I would think that eastern lowland gorillas have about seen it all. As notoriously shy creatures, imagine a habitat as violent as theirs. With humans, a genetic second cousin, slaughtering each other by the thousands and deforesting the rich landscape of the Democratic Republic of the Congo, eastern lowland gorillas couldn't have worse neighbors. Being squeezed into a shrinking homeland, trapped and slaughtered for meat or trophies, or killed protecting their babies from illegal trade, these gorillas of the DRC can't possibly have found genetic evolution to their liking. Nor can they comprehend our mining of their riverbeds for metals used in our ubiquitous cell phones. I'm sure there is a word for coltan mining in gorilla lexicon but I don't think it is freedom of communication. A thirty-year-old dominant silverback can't beat his chest hard enough to ward off this enemy. Further, these remaining 2,500 eastern lowland gorillas don't need cell phones to spread the news among their band that they're in trouble.

No one truly knows what they must be thinking, but it raises the question: What on earth are we thinking?

Humans versus the Planet

There are roughly seven and a half billion people on planet Earth as of December 17, 2019.[2] It is an ongoing debate as to how many humans can inhabit the planet without depleting it permanently. Is it nine billion? Twelve or fourteen? But the question shouldn't be how many people can our planet support but how do we preserve biodiversity and maintain ecological balance for all species? Our successful existence shouldn't be measured by how we have increased the capacity of Earth to accommodate one species. If the pandemic of 2020 has taught us anything it is that we are vulnerable to disease and that dense population proximity is a recipe for contagion. Should we be paying heed globally to what a healthy sustainable habitat looks like? I can tell you it doesn't look like one where a single dominant species has eliminated most others. For one, we aren't worthy of that; for another, it is dangerous. It represents the first line on the jar of ingredients of climate catastrophe. Converting native boreal and tropic forests into cropland to feed twelve billion people will allow greenhouse gases to soar: it will eliminate natural carbon sinks and watersheds, and make freshwater even more scarce, and finally, it will usher in the kind of weather apocalypse we only see in the movies. And, of course, that means that along the way we chose to drop from the ranks our favorite animals, plants, insects, fish, and every other competing species. Global leadership means that we have to monitor human population growth as complex as that topic can be for such a range of distinct regions and cultures. But very few books on ecology and environment that I have read fail to recognize how critical it will be to stabilize population growth. That doesn't mean anyone has totally figured it out however. It would surprise many that we, here in the United States, have not flattened our population growth as of the last census. As of now, it doesn't look as if it is decreasing but we'll see how the 2020 census turns out in the pandemic. Immigration typically impacts the census but we all know that the past few years has been atypical when it comes to immigration. That is stated kinder than the 2020 policy deserves. If that is the recipe here in the United States for leveling out population growth, then we need to search for a sign of a heartbeat.

In contrast to a rising human population, there are approximately 500,000 elephants left in the wild.[3] No one knows for sure how many are left but the ones that remain are endangered, poached, and pinned in. With climate change creating uninhabitable conditions in Africa, where most of the elephants live, it is a desperate fight for survival. As human refugees flee from drought conditions, arable land is going to be scarce for large pachyderms.

Extinction is ugly when it becomes a matter of physics and mathematics. How much land does a stable population of elephants require? We're rounding that corner with a lot of species. Of course money is the common denominator behind the illegal trade in endangered species.[4] Elephants are the largest land mammal and as a result require more resources than most—as much as four hundred pounds of food and forty-five gallons of water a day.[5] Their numbers are not sustainable in a continent where human population continues to mushroom. And it's not a fair fight.

If there ever were an example of this, consider that critically endangered black rhinos are being dehorned in a desperate act to save the species from poaching. With a dangerously low figure of five thousand remaining in the wild, black rhinos are poached for their horns, which are sold for medicinal purposes or more accurately as a snake-oil elixir for whatever ails you in China.[6] Elephants clinging to life must also survive well-armed ivory hunters who have tipped the balance between life and death for their species. And unlike rhino horns, elephant tusks do not grow back. But why can't we just stop buying ivory whether it meets a loophole in international trade regulations or not?[7] It has no medicinal value, it serves no aesthetic value, it doesn't make pianos sound better, and it exploits Native workers. And since we in America imported the most ivory for a century, we have a similar role in the ban of ivory as we do with climate change. We bear the greatest responsibility on a historical per capita basis for global greenhouse gas emissions so we should take responsibility and make amends. We legislated ivory and it's time to do the same with climate change. Beyond reducing my carbon footprint, I'm willing to donate my keratinous fibers to save a black rhino from being poached for the purpose of curing someone's gout. Heck, who knows, a few hundred strands of my hair may prove to be a panacea for quackery.

One of the stories that may not get told in the miasma of a pandemic hangover is that of the fate of minks bred for their pelts. Despite the public outrage over mink and fox farms, the practice is still legal in a variety of nations including the United States. Modern mink farms are like high-end cotton fields except for the heinous crime of animal exploitation, cruelty, and murder. That people worldwide don mink and fox skins in the twenty-first century is almost inconceivable. PETA (People for the Ethical Treatment of Animals) can't decry this practice any louder. I mean, I own a leather jacket so I am guilty of wearing animal skins. Be clear, however, demand for leather goods isn't the reason we slaughter cattle.

Beef, that's what's for dinner, which I can still hear in my head, is primarily why cows are bred. In fact, the demand for hides isn't close to commensurate with the demand for meat.[8] It takes about thirty minks to make one coat, and although I am a fairly large man, it probably didn't take a whole cow to make mine. Getting that off my chest, I find the whole practice disheartening.

The year 2020 proved to be hard on minks, in particular, because of the global pandemic. As I've previously noted, COVID-19 can be transmitted from species to species and to *Homo sapiens*. Minks have caught the virus, as have many other mammals, and because they are factory farmed, they pose a particular threat to workers and the surrounding communities. Minks, like other weasels, ferrets, and otters, are clever and elusive creatures, and often escape from their death row pens and make it to freedom. The added risk of fugitive minks infecting humans was enough in 2020 to call for their mass culling. As of this writing, Denmark, Europe's largest producer of factory fur, has killed a million or more minks and has contemplated the killing of upwards of sixteen million of them.[9] Sweden, Spain, and France have already killed millions more minks in fear of COVID-19 spread. Because the practice is still legal here in the United States, we have followed suit. We may never know how many farmed minks will be gassed because of their beautiful pelt but it does expose the inhumane practice.

We care so little for other living species that we have exposed one to inhumane mass executions. It is time we end animal farm factories, especially one that has no purpose other than the fashion garment industry. This is a business that can't get shut down fast enough. As we lament the loss of millions of humans from COVID-19, please remember that ten times more minks died for their fur.[10]

Threats on Land

The idea that we mine ore for electronic devices including our cell phones in the few remaining habitats of mountain gorillas is yet another egregious example of the destructive capability of *Homo sapiens*. Coltan mining in the Democratic Republic of Congo is taking its toll on gorilla populations.[11] Coltan, or more accurately, columbite-tantalum, is a mineral found throughout the globe but mined almost exclusively in the DRC. Mining efforts in that region can and do initiate conflicts between the species, and poaching is another fateful outcome. Coltan is used in most computers and cell phones for its capacity to transmit an electrical charge without overheating. And although progress has been made to end worker exploitation in the DRC, it continues to this day.

Gorillas have been listed as endangered for decades and remain endangered despite an uptick in population through earnest efforts by numerous wildlife conservation organizations. Their numbers are still precarious, at slightly above a thousand left in the wild.[12] Cell phones and computers outnumber mountain gorillas by the billions. Our family of two (*Homo sapiens* not gorilla) has three phones and three computers in our household. I don't have the faintest idea if any of these contain illegally mined coltan. I know to ask now, and I hope you will join me in asking the same question of your electronics providers. Unintentionally, our blind consumption of coltan may prove to be as wrong as not knowing if a fur coat is real or not.

Few species have a chance in direct battle with economic gain but that doesn't mean we can't make conscious choices. Best-selling author Yuval Noah Harari states in his influential book, *Sapiens: A Brief History of Humankind*: "We [*Homo sapiens*] have the dubious distinction of being the deadliest species in the annals of biology."[13]

Harari details our long history of exterminating species, including the ones competing for the top of the evolutionary mountain thousands of years ago. Gorillas and humans share approximately 98 percent of their DNA; just because gorillas didn't invent cell phones doesn't mean that they aren't painfully aware of their consequences.

Consider the cases of two conservationists in Mexico: Homero Gómez González and Raúl Hernández Romero. Señor Gómez González was bludgeoned to death and then thrown down a well. Señor Hernández Romero was beaten and knifed to death.[14] Their crimes? Preserving monarch butterflies at the El Rosario monarch butterfly sanctuary in Michoacán, Mexico. If we can't even protect the protectors, what possible chance do gorillas, rhinos, elephants, and monarch butterflies have against illegal logging, poaching, habitat destruction, corruption, and land grabs? The deaths of Gómez González and Hernández Romero remain unsolved but considering the fact that the sanctuary resides in one of the most dangerous pockets of the world, it may never be solved.

If we are merely reactive in saving a species once their numbers are drastically reduced, we will become the reluctant arbiters of what lives and dies in the future. If governments cannot sustain regions of biodiversity and establish safety zones for our fellow species, we as a civilization will have begun the plunge into the netherworld. Sr. Homero Gómez González and Sr. Raul Hernández Romero died preserving some of those colors of the rainbow of biodiversity.

Many of our endangered species live in severe habitats, which is a stark reminder that climate change and habitat destruction is a global problem. Its reaches know no limits. Shy of getting into a debate on biogeography and equilibrium theory, which is expertly covered in the classic book *The Song of the Dodo*, by David Quammen, we can agree that habitat destruction is the catalyst for species loss.[15] Changing the atmospheric constitution by burning fossil fuels is equivalent to habitat destruction. Warming ocean temps, rising seas, and melting glaciers have a devastating impact on the Arctic and Antarctic poles. Throw in overfishing, ocean pollution, and encroachment, and havoc can ensue. This is the case with the South Pole and its most distinguished species: penguins. They may be marching right out of natural existence. And excuse me for anthropomorphizing but I bet they would prefer to live on the South Pole versus in your local refrigerated zoo or aquarium. We lose a lot when we care too little.

The Emperor, the King, and the Crookery of Climate Change

As far from us as Earth affords,
we are hunting them nonetheless.
Spheniscidae is unaware of our scheme—
a demonic ploy to radiate the planet.

We feed them krill and squid from fishbowls
and cool them indoors,
thousands of miles from the South Pole
where ice sheets fold into the ocean.

In this era, the age of extinction,
none are excused,
including these well-dressed southerners
from the blackness of deep ocean oblivion.
(2014)

Although I would argue that the red wolf, grizzly bear, whooping crane, and wolverine are worth preserving here in the United States, I am particularly moved by the plight of orangutans elsewhere. The name can be translated from the Malay language as people of the forest. They are not native to the Americas and despite being omnipresent at zoos throughout the United States, they really only have one last habitat for survival: Borneo. There are small families of them on a few of the islands of Indonesia like Sumatra, but their future rests on the largest island in Asia.

Borneo is one of the finest examples of a native rain forest that we have in the world. It is dense with species and diversity, as well as natural resources. Peatlands, which retain moisture and prevent decomposition, are important in the battle to absorb carbon and combat forest fires. The fires that struck Borneo in 2019 hardly received the attention of California or Australia, but losing two million acres of a rain forest like Borneo to fire has devastating long-term consequences. For one, an original rain forest can never replace its loss of biodiversity; secondly, as peatland is converted to a palm oil or tree plantation, it loses its ability to withstand drier conditions that trigger a positive feedback loop just like melting permafrost.[16]

Most of us by now have heard of the controversy over palm oil production. Palm oil is a cheap, highly productive, versatile vegetable oil found in numerous foods and cosmetic products, including Oreo cookies—hint, hint. It also is a source of biofuel for many nations seeking competitive fossil fuel alternatives.[17] I mean, it adds shelf life to processed food, and even prevents ice cream from melting faster. So what's the big problem here? Well, almost all of the world's supply comes from Malaysia and Indonesia, the two nations who claim Borneo. Again, Borneo is the third largest island on the planet and one of its most biodiverse. Burning peatlands for land conversion is a significant source of greenhouse gas emissions. Like melting permafrost, burning rich peat soil releases methane, a far more potent greenhouse gas than CO_2. Transitioning away from carbon sinks like peatland and permafrost is tantamount to firing up your grill with rocket fuel. And those wildfires, like the one in 2019, are raising emissions in the region as well. In 2014, a neighboring nation, Singapore, enacted laws to indict the perpetrators of the air pollution caused by the fires.[18]

Palm oil cultivation and land conversion are not only contributing to climate change, they're making orphans of orangutans. I spent a weekend in Borneo a decade ago and witnessed it firsthand. There are places on the planet where critically endangered species are rehabilitated with the intent of returning them to the wild. I was blessed with the opportunity to witness one of these places.

Samboja, in Borneo, is a dual-purposed village that works to reforest a cultivated region while healing and preparing juvenile female orangutans for reintroduction. Males were also present but quarantined in a separate location; they are territorial, and kept in cages, awaiting release—in effect, prisoners on their own island since competition for land is fierce and palm oil is so valuable. The reintroduction program for the females of Samboja was a six-step process. Each of these steps was on a specific island surrounded by water since orangutans do not swim. My ecolodge room had an overlook that allowed guests to view all but one or two of the islands. We were allowed to tour the island perimeters during guided hours.

Being in the presence of an orangutan in its natural habitat is a once-in-a-lifetime experience. For one thing, their facial gestures are so expressive. One could credibly sense feelings of sadness and boredom, joy and intrigue. It's no coincidence since we share 97 percent of our DNA with them also. Watching their fully functional digits grasp a pole or branch was another reminder that we could shake hands like a next-door neighbor, albeit an orange, hairy one.

My final day at Samboja included a guided trip through the restored forest landscape. Nearing the end of our hike, I was startled by a juvenile orangutan who had been observing us hidden from our view. She approached gingerly like a shy teenager needing to satisfy her curiosity. Locking eyes with mine, she stared right back at me. An encounter between similar creatures with dissimilar fates. It was an exhilarating moment for me. Distilled only by the longer-term realization that her chances of freedom, making a home in the forest, and mating were miniscule; all the while I would be back in my safe and comfy home with my wife in a few days.

People of the Forest

Among the reforested mango trees,

beneath the equatorial sun rays,

a paradise exists in waiting.

It waits

for sanity to emerge

from palm oil plantations,

from corruption, exportation,

exhaustion.

In the storehouse of life,

there exists our future.

In the prospect of hope,

there exists our will.

On the islands of Samboja

deep within Eastern Kalimantan

survives a reticent Robinson Crusoe.

Paprika-dusted hands and legs

glide along a man-made canopy

high atop the trapeze wires

of extinction.

Trading species for teak,

bargaining lives for oil,

carving up a rain forest

is like shortening life's racetrack.

I am too genetically matched

to the subjects I photograph

to not notice,

in the mirror, at Samboja Lodge,

my orangutan shadow.

(2008)

Threats in the Water

Our world's oceans, which cover most of our planet, have not been immune to climate change. Although a good deal of attention has gone to the fact that our large bodies of water have become plastic dump sites, and although it exemplifies a lack of environmental responsibility, it isn't directly related to oceanic climate change. But pollution, in general, is the cause of ocean and air temperature rise. Emitting greenhouse gases is the most egregious example of environmental pollution. Ocean temperatures are on the rise because as carbon sinks, they also absorb excess CO_2. As it does on land, the greater concentration of CO_2 heats it up. Upper ocean surface temperatures reflect the brunt of this concentration of carbon dioxide first, and this also affects the greatest concentration of marine species. Reef colonies are the urban centers of our saltwater world. These coral structures are highly sensitive to changes in water temps.[19]

A few degrees of temperature change destabilize these ocean cities, and all its city dwellers. The balance of the ocean is being tipped. The gradual rise in global temperatures spark the inevitable consequence of sea level rise. The positive feedback of warming poles—basically the melting of sea ice and glaciers exposing areas that would normally reflect the sun—is now absorbing heat and releasing greenhouse gases including methane. Melting permafrost is an almost incomparable threat to mitigating climate change. As is often noted, the Arctic is melting twice as fast as other continents.[20] Warming sea temps and sea level rise are conjoined outcomes of climate change.

An unsettling pall hangs over many of us paying attention to the latest figures. A snowless Northeast may seem quaint but the effects will be catastrophic. Zika virus in Boston, lobstermen following their catch to Canada, sharks off Nova Scotia, the fraying of ocean currents and fisheries, the constant flooding of an aging infrastructure along the Atlantic coastline, fires in the North Pole (Santa, get your turbine hydroplane out), the decimation of the shellfish industry, and well . . . you get the gist of it. Nothing in the end benefits from rising ocean temperatures and sea level rise.

The migration of species due to warmer waters will ripple through our evolutionary chain. We are already seeing some effects of this along the New England coast with sea turtle strandings on Cape Cod and expat lobsters moving to Canada.[21] The proud, generational fishermen of the Northern Atlantic states are facing a very unsure future with or without a pandemic.

As it is now, millions of species are at risk, including the largest of them: whales. Many of these giant mammalian species were once hunted to near extinction. Notwithstanding Japan and a few other nations, whales aren't being hunted by humans anymore.[22] That hardly means that they're not threatened. In fact, North Atlantic right whales may still go extinct in the wild. Many species are either listed as vulnerable or endangered in various ranges of their habitat. Humpbacks, our breathtaking

New England coastal visitors, have made a mild comeback since the whaling moratorium of the 1980s.

Captain Ahab types are no longer a threat, but other threats include ocean traffic congestion with man-made vessels. The tens of millions of container ships bringing those critical, cheap goods from Asia pose as great a threat as whaling ships. Collisions with ships are becoming commonplace.[23,24] The incessant flow of seafaring commerce wasn't an enemy millions of years ago to these giants of the sea. How crowded are the ocean's pathways if whales are crashing into boats? A globalized economy shouldn't ever become the cause of endangerment to whales and their cold-blooded neighbors.

A sighting of a whale is a moment of awe-inspiring consciousness. The enormity of these giant cetaceans is hard to comprehend until you view one from the bow of a boat. A breaching adult humpback is an incomparable sight. Imagine what the first intrepid sailors crossing the vast oceans must have thought witnessing one of these majestic creatures—a slight contrast to modern-day ocean voyagers peering out at the millions of container ships dotting the horizon.

And all these centuries later, we still don't know them well. Studies have found that they communicate for hundreds of miles and have a complex language that may rival humans. Though female humpbacks don't produce the same whale songs as males, the operatic stars of the species, their communication is just as important. Humpback whales have been described as the most elaborate vocalists of the ocean.[25] Their ethereal sounds have been recorded for decades including having made it into vinyl back in the 1970s, when first discovered. Marine conservation efforts since then have been mixed in success depending on the species.

The original crew of Star Trek even theorized that the future of the planet was encrypted in the song of a humpback.[26] Knowing that humpback whales have been spinning tunes for millions of years, the thought that we'd extinguish this species and other cetaceans without listening to them is further indictment of our role as fellow mammalians. Within this magical hidden language underwater is a better understanding of our planet and its plethora of gifts.

Carl Safina, in his recent book, *Becoming Wild*, suggests that we learn more about our own species by better understanding other species and their unique cultures.[27] The oft-used phrase "extinction is forever" is particularly horrifying when you imagine what we might have lost if we allowed the whaling industry to extinguish cetaceans from Earth. Open your ears and listen hard to North Atlantic right whales right now because their songs are fading away in the wild permanently. And although the erstwhile efforts to save the whales have boosted their numbers, they are far from out of the water, to reuse a bad pun.

Humpbacks like other mammals are communal and caring parents. Some species nurse their offspring for years and the females remain in the family unit for life. Sperm whales, like all toothed whales, use a sophisticated sonar system to detect and hunt prey. These clicks are voiced in code, and we're now learning that they signal everything from high alert to personal introductions. And sperm whales produce the loudest sound per decibel of any animal on Earth—creating ocean energy rivaling volcanic activity.[28] Their echolocation clicks have the power to deafen and subdue anything in their path, including free-diving humans studying them for scientific discovery. But we are also discovering that sperm whales choose not to harm them. Sperm whales, massive, big-brained, and shy, display a certain curiosity toward vulnerable human visitors on a free dive. Surely they don't see us as a threat but imagine what they think of us after a few sonar clicks: *Meh, little land bug, if you had our supersized brains, you'd stop polluting this planet.*

Ocean Noise

There are three main threats, outside of ocean warming and acid-ification, to cetaceans: fishing gear entanglements, vessel strikes, and noise pollution. The first two will take some effort to correct in the future but the third is simply inexplicable. Seismic blasting and military sonar experiments pose a real threat to creatures sensitized to noise. Sound by comparison carries four times greater in water than on land. Disrupting a creature's primary source of communication impacts their chances of survival. Author Rebecca Giggs states this in her beautiful debut novel, *Fathoms: The World in the Whale*: "The effects of all this sonic disruption is to shrink the whales' world." She goes on to summarize the effect of ocean noise, "For whales, noise pollution can be lethal."[29] When compet-ing noises invade their habitat, for example, humpbacks struggle to communicate, and in turn get stressed and in peril. In some cases, like that of beaked whales, manmade noise is proving to be even more deadly. Deafening sounds induce them to ascend too quickly, resulting in wanton fatalities.[30]

The military has used sonar at high decibel levels to track submarines and for ocean mapping for years. To their credit, they have become more acutely aware of the threats to whales and other marine species, and they have been using more benign methods of ocean tracking. That hasn't stopped fossil fuel companies from supporting seismic blasting for seeking out oil and gas reserves in the most remote sections of our oceans.

If I were a sperm whale, one of the deepest divers of our waters, seeking prey in the darkest spaces of the ocean, the last thing I'd want to hear is a seismic blast. Their keen ability of echolocating miles below the surface should keep them safe from land's top predator but sadly it does not. They have endured our penchant for violence for centuries but ocean noise pollution borders on blatant disregard. Imagine if an alien species came to colonize Earth but never bothered to study its inhabitants to determine suitability to life on the planet.

Constricting Snakes

I am far from a pure naturalist, and I have an unhealthy fear of long, constricting snakes—thank you, John Winkelman. I wasn't scared of snakes growing up since my neighbor, Jeff Bright, and I hunted them in the overgrown wooded area behind our house. In fact, we once stupidly built a snake pit made up of a variety of local species, and I failed to see the biological problem with that. Back to John though: one day in seventh grade he brought his two boa constrictors to school. Since I had spent time with him at his house, and held his exotic pets, he asked me to help him demonstrate them during class. Everyone in seventh grade is overly self-conscious, and life itself back then was a harrowing show-and-tell experience. As John took one of his boas around the classroom, I was holding the other at the front of class. Snakes are muscle-bound, and boas could be considered the powerlifters of the group. John's boa on my shoulder and arm either got agitated or more likely recognized the fear in its handler and got too close to my neck. Whether it was the air slowly leaving my head or my heightened paranoia, I freaked and screamed for John to get it off me. Of course, I endured the derisive laughter, which was just what I needed for my self-esteem as a thirteen-year-old. A footnote to the story: one of the boas ultimately escaped into John's backyard and was lost for days. John recaptured it but only after a struggle; John was bitten severely and I'm sure bares scars to this day.

Preserving biodiversity is not only our responsibility, it is key to a healthy ecosystem, and yes, that includes constricting snakes. Establishing ecological balance didn't come easily or quickly. The interrelation of species was perfected over millennia, and the balance hasn't been significantly upturned since the dinosaurs died off sixty million years ago. The cause of their demise was ultimately climate related. An extended period of dramatic changes to temperature doomed them to extinction through no fault of their own. If we are currently experiencing a sixth extinction, we are actively causing it. If the dinosaurs taught us something it is that altering climatic conditions can have devastating effect. Having a properly functioning ecosystem supports biodiversity. I would never bet against humankind in its battle for survival in an altered planet Earth but how many other organisms will be with us on the Ark? And where, for the life of us, do we go?

5

CHAPTER

Oh, Sandy   2013

"And Sandy, the aurora is rising behind us. This pier lights our carnival life forever. Oh, love me tonight, for I may never see you again."[1] When Springsteen penned those lyrics four decades ago, an homage to Asbury Park, New Jersey, he had no idea Sandy would be remembered for the storm that took his words so literally and became a reckoning for coastal Atlantic communities. With almost a quarter of the U.S. population affected by her wrath, how do we preserve a way of life for so many? With damages in New Jersey alone totaling more than thirty billion dollars, many commercial establishments will struggle to return. But they must return. Boardwalks and Jersey beaches are a link to intergenerational memories and an economic linchpin for a state with 130 miles of coastline. Superstorm Sandy will spawn other once-in-a-lifetime storms and intense climatic activity will challenge seasonal

industries in the future, but sustainable commercial development in the form of minimizing beach encroachment and erosion, night pollution, energy and water waste, and carbon emissions, to name a few, can become the girders of a sunny, sustainable Jersey Shore. Recognizing that we yearn for the ocean, and that escaping the summer heat of our Eastern cities for the shoreline is an annual rite, we will need to rethink our relationship with nature.

Instead of subsidizing rebuilding efforts with diminishing funds, we must build with Sandy in mind. We've reached an impasse and simply cannot afford to ignore the impact of changing climatic activity. With taxpayer-based funds supporting the federal flood insurance program, now running in the red, we better start using climate predictors in conjunction with architects and insurers. In a time of political paralysis, it would be financially prudent to quit allocating funds for unsustainable building practices. And even if the biggest developers can get flood insurance at a premium, does that afford them the right to gamble on nature? Only Atlantic City would take those odds. Raising sea walls is becoming tantamount to playing Russian roulette with water. We need to respect the tidal push of hurricane waters and build to suit nature. Instead of insurers raising premiums to cover new structures that may not survive storms to come, we need to design with a better understanding of the consequences of rising water and temperatures so the sandcastles of the future don't wash up on Main Street.

Asbury Park, like all Eastern beach communities, won't forget Sandy anytime soon. Coastal developers everywhere shouldn't either. We cannot place hotels and other commercial buildings in the path of the next Sandy. We don't know when the next superstorm will come and level beaches along a thousand-mile swath; we just know that it will come. Let's be ready next time. Let's allow nature to emote without tragedy. The Jersey Shore, for one, deserves it. If only Madame Marie were still around.

Climate Change and Superstorm Sandy

The consequences of climate change are many. I have highlighted a number of them throughout this book. And though there are armchair experts that equate weather to climate, they are distinct. Climate is defined by long-term patterns of weather and does not reflect just a single year, a record storm, or a severe summer drought. Changes in climate are most assuredly contributing to weather anomalies but they shouldn't be conflated, which is a common occurrence. Recalcitrant politicians make erroneous statements about climate change on extreme cold days in the winter all the time. When a politician brings snowballs into the halls of Congress on a record cold day in the South to bolster the denial of climate change, it demonstrates a fundamental lack of comprehension. The patterns of climate comprise weather conditions over an extended period of time. One incident doesn't make a scintilla of impact on the overall climate. There will be warm days in the winter, cold days in the summer, forever. The impact of climate change that most of us have witnessed firsthand is an intensification of weather-related catastrophes. Most of us have been affected by a wildfire, hurricane, superstorm, tornado, bomb cyclone, drought, or flood. And the quantity of these events hasn't changed significantly, but the intensity of them sure has changed.[2]

With deference to Katrina, Irma, Maria, Harvey, Dorian, and other recent hurricanes, I wrote about Sandy for a reason. Sandy was labeled a superstorm, and it shook people in the Northeast to its core. As ocean temperatures increase, and as the air gets more humid, conditions can be ripe for a superstorm like Sandy, the effects of which are still being felt. Sandy blasted coastlines from Virginia to Connecticut, sinking subways and urban streets. It sliced off parts of the Jersey Shore. Homes were lost; lives were lost. Commercial and residential properties were permanently damaged. Bruce Springsteen's beloved Asbury Park took a near direct hit. Madam Marie, officially Marie Castello, wasn't around to predict Superstorm Sandy; she passed away in 2008 at the ripe old age of ninety-three.

The New Jersey McGraths

I married into a Jersey family—they're not all like the one with Snooki, for the record.[3] My wife Maryann's family has called New Jersey home for generations. Her fondest memories include basking in the sun on one of countless Jersey beaches that stretch from Sandy Hook to Cape May. It was a summer retreat taken by many families including her twenty-three first cousins. One of which, Pam Klink, recalls feasting on crumb buns at Kay's Bakery and miniature golf at Barnacle Bills. Congregations of kids in bathing suits all day, lazing in the surf, and playing pinball at Peek-a-Boo convenience store. And, hazy, warm summer evenings that included Wednesday night fireworks and Little League baseball games under the lights at Lavallette. Pam and her four siblings netted crabs in the shallow water beneath the piers or haul in fish from the soft sandy beaches. Not only was it a tradition of her family, it was a virtual rite of Jersey citizenship. Maryann recounts being sardined with her siblings into a four-door sedan that lugged behind a trailer jammed with beach gear. This included the madness of beach traffic on the Garden State Parkway with just about every other Jersey family on the same pilgrimage. As with most coastal communities, some beach bungalows are seasonal rentals, while others are year-round residences, not that nature makes a distinction.

The fall of 2012 became a reminder of, and referendum on, climate change. Was climate change to blame or was it a freak weather condition that caused it? Likely, it was a combination of both. But that also means that it could and will happen again since the instigating factors have not diminished in any way. Hurricane season isn't just for Florida and Louisiana anymore.

In 1988, Pam's parents and Maryann's aunt and uncle, Bob and Doris McGrath, built a year-round home in Ortley Beach, on the original site of Doris's family's cottage. Ortley Beach, New Jersey is bookended by the bay and the Atlantic on that long strip of land that makes up this unique peninsula. They had spent many summer days and weeks at the tiny cottage Doris's parents built back in the 1940s before becoming permanent residents; that is until Superstorm Sandy struck the coast with such ferocity in 2012.

Bob McGrath served in World War II, as did his three older brothers, one of whom, Bill, was shot down in the Philippines, and went missing for forty-four days. The compelling story of the McGraths is the converse of the fictional Ryan brothers of the movie *Saving Private Ryan*.[4] Bob, the youngest, went to the extent of lying about his age to a recruiter knowing he didn't have his mother's consent. What mother would want to send off all her sons to distant battlefields across Europe and Asia? Bob survived the war, as did all the McGrath boys despite some of the harrowing experiences they all endured. Doris's family wasn't so lucky. Her brother, Buddy, a tail gunner on the B-29 *Arkansas Traveler*, was killed in action in 1945, less than three months before the end of the war in the Pacific. He was eighteen. Kenneth McGrath, my father-in-law, was the first to serve, and the lone Marine. He was sent to the Pacific and in 1945 personally witnessed the surrender of more than half a million Japanese troops in Tientsin, China. Ken was closest to Bob in life and in age. He passed away in April 2011, long after the two older McGrath boys had died, and he rarely spoke of his service. According to Pam, her father, Bob, didn't either.

Not that Superstorm Sandy needed any assistance in wrecking Jersey families, it took a heavy toll on Bob and Doris. These two, shaped by the Second World War, parents of five children, were in the twilight of their lives when Sandy arrived in 2012. They had suffered hurricanes and storms on the Jersey Coast for fifty years, and the barrier islands despite being highly exposed to Mother

Nature had mostly held firm. But this storm was unlike anything they had experienced. That anyone had experienced there.

Water temperatures in the Atlantic were unseasonably high in 2012, a condition ripe for hurricanes. Coupled with a blast of cool arctic air, a separate storm pushing its way inland, and the tidal impacts of a full moon, Sandy was the perfect storm.[5,6] This fusion of meteorological activity confounded storm predictors and made landfall hard to pinpoint. Dr. Jennifer Francis, a meteorological scientist at the Woodwell Climate Research Center, says that the warming arctic air that weakens upper-level winds makes waves move much slower and helps storms linger for greater periods of time.[7] Once this arctic air mixed with the approaching category 1 hurricane in the Atlantic, Superstorm Sandy was officially born. And she was a massive one-thousand-mile-wide baby hurricane that stalled over the Mid-Atlantic Coast for days.

A hurricane rarely defies the easterly pattern of the jet stream that typically spares this region from the direct impacts of Florida or the Carolinas, for example. Sandy changed all that. And although most of New Jersey got struck by Sandy, Ortley Beach, which is officially part of Toms River township, was hit particularly hard, with hundreds of homes disappearing altogether. Locals began referring to it as "ground zero."[8] Ocean County in New Jersey offers up plentiful coastlines but sustained rainfall and tidal surges also expose it, which is exactly what Sandy did. With record smashing effect, Sandy became the strongest tropical storm to hit the Northeastern United States, and only Hurricane Katrina in 2005 had been as costly.[9]

I had never heard of a superstorm before Sandy but be assured we'll hear more of them moving forward. As arctic temperatures continue to rise faster than anywhere else on the planet, the disruption of the polar vortex is causing arctic air to disperse to areas beyond historical patterns.[10] This phenomenon is wreaking havoc in the United States in 2020, in the form of May snow showers and record rainfall in the Midwest.

It may seem a stretch to suggest that climate change was a factor in Bob and Doris's displacement but what else would you call it? Their home for twenty-four years was destroyed beyond repair. According to Pam, they packed for five days in advance of Sandy, and never returned. The damage from the storm to the barrier islands was devastating and longstanding. By the time, their sons, Peter and Paul, returned to the house, three weeks had passed. Access to the area was limited and they were shuttled to and from Ortley Beach for two hours at a time. The home was uninhabitable and most of its items beyond salvage, including Doris's satin wedding gown from her mother-in-law's famous dress shop in Plainfield. Irreplaceable items from two lives well-lived lay in ruin. To add to the tragedy, Doris, at eighty-three, was already in physical pain from a fall when she learned that her last remaining sibling, who lived in South Carolina, had died. Pam added that her folks were always comforted by their deep faith in God, which helped them in times of crisis. This period would test their resolve.

Uncertainty had become a new condition of life for many citizens of the Garden State. As realities set in and the challenges of recovery and restoration became apparent, gaining access to the property and all its contents was virtually impossible.

Peter remembers Bob saying to him, "Peter, the only thing that matters to me is that you retrieve my dog tags." They not only found them but Buddy's Purple Heart medal as well. Despite the fact that their home was on a five-foot foundation on property closer to the center of the island, the surge eclipsed it by almost a foot. Due to the extensive damage, the recovery process was a drawn-out battle. Utility and infrastructure repairs took a year or more; for older homeowners, the thought of rebuilding was unrealistic.

Bob and Doris were not unaware of the risks of living in Ortley Beach; with both flood insurance policies and architectural recommendations for tidal surge, they felt safe when they built in the late 1980s. Superstorm Sandy changed all that. They fully expected to return to Ortley Beach once the storm passed. In the end, they were able to salvage the foundation, and in a way, they were fortunate to have been properly insured. They received insurance payments as well as Federal Emergency Management Agency (FEMA) checks. According to Pam, her folks had the financial compensation to rebuild the house but the area wasn't close to restoring basic utilities so the chances of them returning at that stage in life were low.

Flood Insurance and Rebuilding

Many families were unable to rebuild their homes even after flood insurance settlements, some of which were grossly insufficient. Others simply chose not to rebuild due to the potential of a future Sandy. Coastal properties have a more recent concern: Should they be built or rebuilt knowing that climate change will affect further erosion of coastal lands? It is estimated that New Jersey has spent a billion dollars over a thirty-year period to restore its beaches.[11] The yearly economic game will be to measure tourism dollars versus coastal repair in the face of increasing sea level rise. I would never suggest that New Jersey shouldn't support its beaches and its summer visitors, especially if I want to be invited back to any family reunions, but rebuilding without assessing the long-term effects of climate change makes no sense to me. Private insurers aren't going to lose money either, by the way; they study climate change closely.[12]

Some companies even choose to negotiate in bad faith regardless of the suffering of property owners. There are pending lawsuits still from Sandy victims who have never recovered losses despite carrying supplemental flood insurance.[13] Since FEMA only provides a moderate amount of coverage per the NFIP (National Flood Insurance Program) guidelines, many wealthier property owners in higher-risk zones seek private flood insurance as supplemental backup. Not everyone has the financial wherewithal to do that, however. As the region gets rezoned for base flood elevation, it will get more and more expensive to insure property.

My brother-in-law, Kevin McGrath (Ken's oldest son), who spent his entire work career in the insurance industry, doesn't think that the U.S. government should be in the business of funding flood insurance for home and business owners who build in high hazard zones to begin with. He went on to say, "If someone wants to build a home or hotel on a beach, they should go buy private flood insurance at Lloyds of London; they may think twice about building on the beach when they see what it costs in the private insurance market." Sandy forced FEMA to quickly reassess flood zones as a result of future exposures. Kevin called Superstorm Sandy a "game changer" to FEMA. And there always was a certain risk living close to the water anyhow, but how does one prepare for the growing impact of climate change? How high is enough for the next Sandy? How about six inches or more every fifteen or twenty years? Sealevelrise.org estimates that sea level on the East Coast will rise between ten and twenty inches by 2050.[14] Welcome to second-floor living.

We think of African farmers struck by drought being forced to flee their homes when we think of climate refugees; we don't think of Bob and Doris McGrath of Ortley Beach, New Jersey. After being shuttled back and forth from New Jersey to Pam's home in New Hampshire, it wasn't long thereafter that Bob passed away. Who knows if Superstorm Sandy played a role, but it doesn't change the fact that losing a home is a severe personal loss at any age. Being upended in retirement is hard on anyone but as a consequence of hurricanes, tornadoes, wildfires, it is also unjust. Bob and Doris's middle daughter, Linda, described it this way: "My parents were shattered, decimated. Overnight their life was shattered. They thought they were going away for a weekend. Everything they knew, their home, their community, their life, was gone. Life as they knew it would never be the same. It is so hard to put into words the emotional and physical trauma that was experienced." Peter put it more bluntly, "You're never too old to get kicked in the crotch." Doris bravely carried on for another six years—mostly in New Hampshire. She succumbed to her illnesses in October 2018, at eighty-nine. One tough lady.

The Boss

Sometimes it seems to me that Bruce Springsteen has the weight of New Jersey on his shoulders. What other state has a more prominent native son? It couldn't have made it any easier when the most destructive storm ever to hit the Garden State shared the same name as one of his hit songs. He has spoken for many New Jerseyans through a half century of music. The lyrics to his songs are akin to biblical verses for his diehard fans. My first exposure to The Boss was in 1975. I purchased *Born to Run*, arguably his most famous album, after reading Mom and Dad's copy of *Time* that featured him on the cover.[15] It was transforming for me, as it was for so many in my youth. I listened to that album spin on the turntable with headphones on for years. I could recite all of the lyrics in my sleep. For a rebellious teenager, his music unleashed an abundance of emotion and angst in me. Not that I needed additional encouragement to be moody, distant, and delusional. For a safe and sheltered, middle-class child of loving parents, I had to search hard for external strife. Still, the teens sucked and without school athletics, I might have just spent too much time beneath that giant Exxon sign. My only sibling shares his first name with Springsteen and I'm never sure which Bruce my wife is referring to when the name comes into conversation. Even today, we spin Bruce's forty-fives (you know which one) on a restored 1963 Wurlitzer jukebox that is just below a large framed image of Bruce (still not my brother) and Clarence (E Street Band saxophonist). In 2012, his songs from the *Rising* inspired by the 9/11 terrorist attacks provided comfort again for fans during Superstorm Sandy. We look to leaders in times of crisis. Springsteen may be a reluctant leader but he cannot escape his influence on much more than just rock music. As expected, he chose to pitch in to assist his stricken home state. The charity performance, Concert for Sandy Relief, on December 12, 2012 raised close to fifty million dollars.[16] It's worth noting that he was but one of the many performers that night. In fact, it would be easier to list those who didn't perform. Bruce partnered up with another famous New Jerseyan, Jon Bon Jovi, for several of their biggest hits, including one of Jersey's anthems "*Born to Run.*" Springsteen has given back to his home state in abundance. His relief efforts, and his collaboration with Governor Christie, proved to be critical to New Jersey post-Sandy. It isn't hard to see why he is revered there. There is some truth to music being the universal language. When people of all stripes can band together behind music to lift up others in need, we rebuild faith not just cities. Springsteen has done his part.

For the great state of New Jersey, 200,000 homes were damaged and over a hundred lives were lost.[17] For many, their hopes and dreams also washed away under Superstorm Sandy. That is the ugly truth about climate change. The intensity of weather will affect how we live our lives along the world's coasts and pretty much everywhere else. Can we design for the future if we don't prepare for the next weather-related catastrophe? If you aren't asking that of a developer, builder, and designer in the future, it will be a lesson lost from Sandy. Sea level rise is a fact of the present and the future; living with it is more than resilience. We must design homes that are both green and resilient, and we need to make them as geographically appropriate and as energy efficient as possible. There are excellent green building programs available to businesses and homeowners—much better than in 2013—so there is no excuse for designing in the past.

My friend, Tim Hughes, of C&S Companies in New York State, an experienced engineer specializing in green buildings and energy efficiency, says that there are a number of excellent programs including BREEAM (Building Research Establishment Environmental Assessment Method), LEED (Leadership in Energy and Environmental Design), Living Building Challenge, Passive House US, the WELL Building Standard.[18] He, like me, is passionate about green building practices, and about creating our built environment in far more sustainable ways. If we fully comprehend the economic effect of Superstorm Sandy, we really don't have a choice. And if the lessons from climate disasters are heeded, then we move ahead on a singular path. One that meanders down the Jersey Coast with thriving beaches and boardwalks where young and old share crumb buns and Kohr's custard. A coastline prepared for Sandy's offspring with homes and businesses filled with memories not floodwaters. Summers that allow tourists to gamble with tokens not lives; summers that safely seduce voyagers to the Stone Pony for years to come, and summers that create memories not havoc. Perhaps when you visit, you'll brush by a family whose lives were enriched by that hundred-mile stretch of land for generations. One just like the McGraths.

Oh, and I love you, Bruce (my brother).

6

CHAPTER

The One-Way Flight of the Danaus Plexippus   2014

*O*n this forty-fourth anniversary of the first Earth Day, I wanted to recognize the state of the *Danaus plexippus*, or as we more commonly refer to them, monarch butterflies. Migratory creatures have always fascinated me but few as much as the monarch, which migrates like songbirds. Lately they have been appearing in the news for the wrong reason. They've earned attention because of their drastic decline in numbers.

Their annual cycle of migration is one that requires the transformation of four generations of the species. The two-thousand-mile or more journey of the monarch butterfly from North America to Mexico is one of storybook fantasy. An inner compass steers them to familiar plots of plants year after year much like sea turtle migratory behavior. But their compass has been compromised of late by various sources, one of which may be anthropogenic climate change.

Imagine if your chance at becoming a parent was reliant on a very specific set of factors: one region, one plant, one climate. The ecological balance of species survival is so fragile, and we are tipping the scale for these majestic butterflies. For the monarchs, subtle changes disrupt their flight pattern, their food supply, and their eggs.

Climate change may not be reversible, but the culling of milkweed plants certainly is. Although Mexico has made strides to reduce impact on this butterfly's Southern habitat, we in the North have not. Our incessant need to eradicate weeds, including milkweed, may trigger the end of a variety of native species, including monarchs. It may also be responsible for the alarming loss of all of our pollinators in the United States, not just Danaus plexippus.

Monarchs have been on the decline for a decade now. I'm far from an entomologist, but I have firsthand knowledge of this. My uncle, Bob Clark, who owns the family farm in Upstate New York, decided years ago to allow the growth of milkweed on a former grazing field because they enjoyed seeing the monarchs every year. Back then he told me that his field was full of butterflies and that he was going to keep it for that purpose. Last week at a family get-together, I asked him if he still had the field and if they had continued to come back. He just shook his head. "Hardly any," he said. "I barely saw any last summer and saw almost no signs of chrysalis [pupa stage] on the leaves." His experience is similar to others. According to official reports their decline in Mexico has reached an alarming state. The occupied area of monarchs has shrunk by over 50 percent in just two years.[1]

Nature never ceases to amaze me. Of the millions of species that inhabit our planet, the Danaus plexippus has to be one of the most colorful. As a kid I loved seeing monarch butterflies chase my back throughout the summer. Increasingly we are finding that relying on single crop production has negative consequences, which suggests we should revisit traditional agricultural methods that support heterogeneous fields. My request for this anniversary of Earth Day is for those of us who can to plant some milkweed and put away the herbicides. After all, without species diversity, it becomes very black and white.

Monarchs and Milkweed

When you fully grasp the migratory life of monarch butterflies, you cannot help but become a fan—perhaps even become someone who will plant milkweed in any stretch of untended land, and who roots for their journey to continue endlessly. Someone who might even pick up an organic pesticide to tend to their lawns, gardens, and crops. We do not have the most desirable backyard in the neighborhood, to say the least, but we get these colorful nomadic visitors every year now.

Monarchs view the world from eyes with thousands of lenses that not only see the spectrum of colors but ultraviolet light as well. These characteristics aid in their migration. It is well known that monarchs have a unique connection to milkweed especially for successful reproduction. It is the only host plant for eggs and larvae; that's a lot of faith in one plant. Milkweed is truly a weed, and if you choose not to deadhead it, it presents a challenge. Of course we don't deadhead it, so from the sticky latex that is akin to holding a melted Cinnabon Classic Roll on a hot day to the seed hair that looks like something Tim Burton would invent, it is a messy plant. We haven't experienced any decline in milkweed production despite my inattentiveness (think laziness) so our little weed garden experiment has been a success.

The attention surrounding the steady decline of monarch butterflies has proven to be critical to their chances of survival. Allowing milkweed to grow freely in some parts of the United States has been providential to monarchs. According to recent studies, their numbers have steadied in the Eastern United States, while in large agricultural regions of the country their numbers continue to plummet.[2] Modern mass farming practices simply do not allow milkweed to coexist with prime food crops.

Although I can commiserate with monarch larvae about having a limited diet, I can actually eat other foods. They subsist on milkweed leaves alone, which is also a survival mechanism since it makes them less tasty to other predators. When those weeds are eradicated in our nation's breadbasket and in our fertile West Coast valleys, the cycle of life for the monarch comes to an end. Surveys show that their numbers in the West have fallen hard.[3]

We typically think of large avian species like terns and geese when we think of aerial migrants. Smaller species like hummingbirds and sandpipers also make impressive journeys, but monarchs are aeronauts of the insect kind. Their trips take them across international borders through a multigenerational experience. Their wintering spots in Southwestern Mexico are revered by the many who come to witness this spectacle. Only nature can provide these spectacular moments, an event much like olive ridley arribadas, which is why monarchs hold a special place in many of our hearts.

I remember one time a few years ago that I spotted a monarch and wondered when I had last seen one. Where had they been? I was concerned about their drastic decline in population and I spoke of them fading out into permanence—not too subtle reference to extinction—in a poem from 2007. Dane St. Beach is one of the popular local beaches here on the North Shore of Massachusetts.

Dane St. Beach

Retirees pass in front of me,
picnic baskets, folding chairs,
drawn to the pure blue sky
that trails a storm.

Sailboats at anchor,
like plovers in the current,
are pirouetted by
full masts at eight knots.

A monarch butterfly glides by
then flickers like tail lights,
a spectrum of nature's color
fading out into permanence.

Swirls of high tide caps
smash into embankments
along the horseshoe beach-
moats to the castles of Beverly.

Middle-aged men sit cross-legged,
playing chess and hooky.
An elderly man in a blue jacket
totters behind his black mixed–breed.

My shoulders still amber

from August's Carolina wedding

sap the sun's nectar,

the final syrup of summer.

I should finish my article,

answer my e-mails.

For heaven's sake,

what am I doing here?

But I slide back into the bench

at post above Dane Beach,

angle myself in homage to the sun

and release this idle mind.

(2007)

Glyphosate, Genetic Engineering, and Corporate Agriculture

Glyphosate is the active ingredient in Roundup. A chemist from Monsanto discovered that it had the properties of an effective herbicide in the early 1970s. Since its introduction in 1974, Roundup has become the most commonly used weed killer in the United States.[4] Roundup will kill weeds indiscriminately, so its widespread usage includes monarch larvae's favorite food: milkweed. The eradication of milkweed through applying glyphosate severely impacts monarch butterfly habitat. The successful journey of monarchs is highly dependent on milkweed, plain and simple. Roundup is a broadly recognizable brand, not so much the chemical, glyphosate.

Forty-five years after Monsanto introduced Roundup, it is in the news more now for its potential lethality to humans. This is expertly detailed in the award-winning book, *Whitewash: The Story of a Weed Killer, Cancer, and the Corruption of Science* by Carey Gillam. She writes, "Many researchers fear that one of the worst impacts of glyphosate on human health may be as an endocrine disruptor, a dreaded term for chemicals that interfere with hormones in the human body in ways that can cause cancerous tumors, birth defects, and other developmental disorders."[5] The author is a veteran journalist who has covered corporate agriculture for twenty-five years. She has long covered Monsanto and has met with many of their scientific experts over the years to determine the safety level of their products. Throughout the book, she exposes the company's massive efforts to counter any alarming data or negative press. If you think the EPA can't be subjected to corporate influence, wake up. As a result of her investigative work and additional scientific study, farmers and gardeners have been successfully suing Monsanto after contracting cancer, primarily non-Hodgkin's lymphoma—a disease our family knows firsthand.[6]

Many consequential lawsuits have been filed against Monsanto but that crime against humanity wasn't enough to deter Bayer AG from purchasing Monsanto in 2018. Glyphosate is partially banned or severely restricted in many other parts of the globe, including France, Italy, Luxembourg, Germany (oh, the irony), Greece, Spain, Colombia, Mexico, Saudi Arabia, Oman, India, and Vietnam.[7] If more nations cease from importing it, it could prove to be a real blow to its market dominance. Here in the United States, we have no ban and it can be found prominently at our big home improvement chain stores. You don't have to be too imaginative to know which ones I'm talking about. Progress is being made slowly here in the United States. Los Angeles County has banned its use on county properties, for example.

Roundup alone may not even be as impactful as the development of Roundup Ready seeds, which are genetically modified versions of the conventional seed. I often refer to these seeds as "Frankenseeds" since they weren't intended to be what they turned out to be. Through this patented new seed, farmers can plant giant monocrop farms ostensibly to contain weeds and increase productivity. These modified seeds are resistant to the chemical glyphosate and are by no means your average garden store variety seed. They are controlled by large agrochemical corporations: ChemChina/Syngenta, Bayer/Monsanto, BASF, and Dow/DuPont, now called Corteva Agriscience.[8] All are multibillion-dollar corporations. Interestingly enough, only one of them is now a U.S.-based company. These Frankenseeds withstand the onslaught of Roundup to eliminate conventional weeding and make your corn on the cob a healthy summer picnic standby. Farmers have also learned the hard way that to simply replant these Frankenseeds is in violation of the chemical company's patent rights.[9,10] Airborne Frankenseeds that make their way to another's land, by the way, can potentially ruin organic crops for years. It is but another draconian measure to monopolize the farming process.

No small farmer has even a faint chance of staying the legal course against these Big Ag-Chem corporations. Farmers have also discovered that once in bed with these seed sources, it is difficult to get out. As if small farmers needed any more challenges to their existence, being indentured to Monsanto practically sealed their fate. Large crops that have been planted from genetically modified seed and sprayed glyphosate don't easily revert back to fertile farmland either. It is has been established that these chemically active crops negatively impact soil productivity.[11-13] Just like bacteria and antibiotics, weeds grow resistant to repeated chemical applications. Farmers take to dumping more glyphosate on their farms to eliminate weeds with diminishing results. There has been debate as to how effective these Frankenseeds are in the first place. These are not universal heirloom seeds by any means. As the reliance on Roundup increases, the land becomes less fertile and more and more toxic. Over time these farmlands become dead zones. All the while, their seed producer benefits.

Most folks have reached the same conclusion that I have regarding the food chain: I'm no chemist but I'm pretty sure that I don't want to be eating this radioactive crap, nor do you. Modern commercial farming doesn't appear to be much of a symbiotic relationship. We have reached this point because we don't farm like we did once as an agrarian nation. We grow fewer crops, just more en masse. Despite the advance of measures like no till farming, which strengthens topsoil, we are still heavily reliant on chemicals. A whole lot of them.

Traditional farming methods using rotational crops and diversity need to make their way back to American agriculture—methods that allow for milkweed patches and monarchs. Small, local farms can prosper again if we truly equate food to health. The more we rely on native and organic methods of food production the better our well-being—statistics bear that out.[14] America's breadbasket is primarily in the monocrop business. Most of our corn, soybeans, and wheat come from about five states. These comprise the heartland of America.

It is hard to comprehend that America produces enough grain to feed much of the world. That is if it wasn't already appropriated for the beef industry. America is the largest beef producer in the world, outproducing all of the European Union, and soy and corn are the beef industry's prime feed grains.[15] Those monocrops carefully doused in herbicides and pesticides go directly into our food supply. So, it isn't just the pollinators like butterflies, birds, and honey bees that are getting sick. When our children are heavier and suffering higher instances of diabetes and asthma, something is happening. America is showing signs of a sicker populace despite the fact that smoking has declined considerably. Never has the truism "we are what we eat" been more apropos.

I remember watching a documentary a decade or more ago called *Genetic Roulette* that made the claim that genetically modified foods were a factor in childhood obesity.[16] In my simple mind I could grasp the point that eating animals fed genetically modified or GMO grains to make them plumper faster might not be a good thing for our kids, based on obesity statistics.[17] I'm sure there were a million articles funded by Big Ag corporations disputing this conclusion but it stuck with me.

I dedicated chapter thirteen in *One Green Deed Spawns Another* to food and its connection to health and democracy through the lens of the inspiring Frances Moore Lappé. Go back and read her bestselling book, *Diet for a Small Planet*, again. Fifty years ago, she showed all of us how to eat healthier diets.[18] She and I have had discussions because it hasn't gotten easier to find out what's in your food. In 2020, we still have food labels that do not disclose if GMOs were used in the product. That puts a lot of burden on the consumer.

The easiest way I can determine the existence of GMO grain in food products is to look for cornstarch, corn syrup, corn oil, soybean oil, canola oil, or granulated sugar on the label. So, just think how much GMO grain goes into your body if you're just casually shopping at your grocery store. Frankie, through her work at Small Planet Institute, has been educating consumers about modern corporate agricultural practices, including the fight for greater transparency and the negative impact of monocrop farming.

And lately even the scientific community has gotten more vocal. When scientists have to band together to petition for the ban of a chemical, we should probably listen. This effort to ban glyphosate wasn't done on behalf of monarch butterflies but they are clearly a beneficiary also. Eighty scientists wrote and signed the following petition regarding glyphosate:

The World Health Organization's recent reclassification of glyphosate as a "probable human carcinogen" is only a small part of the known toxicity of glyphosate herbicides.

Chronic exposure to glyphosate herbicides is associated not only with cancers, but also with infertility, impotence, abortions, birth defects, neurotoxicity, hormonal disruption, immune reactions, an unnamed fatal kidney disease, chronic diarrhoea, autism and other ailments.

In addition to human diseases, glyphosate herbicides are linked to more than 40 new and re-emerging major crop diseases. They are causing irreparable harm to the entire food web; including the plant kingdom, beneficial microbes that supply nutrients to our crops and soils, fish and other aquatic life, amphibians, butterflies, bees, birds, mammals, and the human microbiome.

For the sake of the planet, our children and our grandchildren, all spraying of glyphosate herbicides should be immediately replaced with eco-friendly alternatives that restore damaged food webs. We urge you to have the courage to stop the destruction of life on our planet as leaders for future generations.[19]

Chemical Titans

It aggravates me that we have to wait until a chemical, like glyphosate, proves to be carcinogenic before we cease its use. Perhaps we should pay attention to dying monarchs, honey bees, or hummingbirds before it sickens humans too. We often find out after the damage is done that there were risks that a chemical company was willing to take, and we see it over and over again. DDT, nerve gas, asbestos, phthalates (plasticizers), PFOA (perfluorooctanoic acid), CFCs (chlorofluorocarbons), fire retardants, mercury compounds, PCBs (polychlorinated biphenyls) and the list goes on and on and on.

Commercialization of potentially toxic chemicals is becoming a new white- collar crime. Some chemical companies have determined that the nominal EPA fines are just a factor of doing business in the industry. The history of chemical giants is dotted with examples of corporate greed over consumer health. Just as Carey Gillam exposed Monsanto for its dirty deeds, the long, sad and sordid history of Teflon comes to mind also. Detailed in the 2017 documentary *The Devil We Know* and the 2019 film *Dark Waters*, the chemical compound, PFOA, used in Teflon and other nonstick surfaces was found to cause multiple cancers, thyroid disorders, ulcerative colitis, and birth defects.[20,21] Turns out that DuPont and 3M's super chemical that coated nonstick surfaces, and waterproofed your sofa didn't break down under fire or water, or anything else for that matter. That may seem like better living through chemistry but when it was discovered that almost every human on Earth carried this chemical in their blood, it was shocking.[22] Although DuPont knew of the potential harm of the product for many decades, it served its shareholders first and foremost.

We are increasingly aware that some corporate chemical titans have played free and loose with chemicals and risk. It took an international agreement in 1986 to ban CFCs, commonly known as aerosol propellants, from further destroying a lifesaving atmospheric layer; it took seventy years to ban TEL (tetraethyl lead) the additive in leaded gasoline, despite knowing of its toxicity all along.[23] I was stunned to see it still being offered at gas stations in Indonesia when I was there in 2008.

We have to hold chemical conglomerates responsible for their products just like any other company or individual. With environmental cancers like non-Hodgkin's lymphoma on the rise, we may want to reassess how chemicals make it to market. The U.S. government has been tardy in banning chemicals found hazardous by many other nations. Not sure how we built up immunity to glyphosate here in America but our government apparently thinks we did.

That's enough to convince me that the cautionary principle should be, and should have always been, in place when we're talking about our food for heaven's sake. Remember the names: Monsanto, Syngenta, Dow Chemical, and DuPont. We may not be calling them by those names today but we won't soon forget the damage done.

And recall the story of butterfly activist, Homero Gómez González, from chapter 4. If there is any question as to the dangers in standing up for our fellow species, let's never forget him.[24] As one of the most prominent butterfly activists in Mexico, Gómez González managed the El Rosario Monarch Butterfly Preserve until January 2020 when he was found floating in a well. His efforts to ban illegal logging and conserve vital land for breeding monarchs put him in danger. A former logger himself, Homero grasped the potential of ecotourism and monarch butterflies. He was ultimately successful in getting logging restricted in a region of Michoacán, which pitted him against the logging community, drug cartels, and avocado farmers. In the end, it cost him dearly. Homero Gómez González died trying to save monarchs; he was fifty.

Loss of Pollinators

Most of us have heard about the threats to our pollinators and food supply. And though no one would refer to monarchs as top pollinators, they are but one of many that keep our flora productive. The one highly effective pollinator that has seen declining numbers is the honey bee. I know that we don't see many honey bees in our flower or weed gardens anymore. And the idea that an insect called "murder hornet" has made its way to our shores to devour honey bees is bad news even in the era of the deadly coronavirus. My family can tell you that I'm already afraid of large flying insects. The sight of a two-inch bee sporting a Halloween mask would keep me quarantined for the entire summer with or without a novel virus in the air.

Without our intrusion, honey bees would be stable in numbers and hard at work pollinating and supporting their nest and queen. They are the top pollinator of our food crops, and a sign of a healthy ecosystem. By the way, a female honey bee sacrifices its life when it stings a human. A "murder hornet" is a bit more ghoulish, having no limits to its infliction of pain. Most bees and other insects are pollinators but honey bees are clearly the champs. It is estimated that they are responsible for pollinating a third or more of our food crops; losing the honey bee population would be catastrophic to our food supply chain.[25]

We began to hear the term *colony collapse disorder* a few years ago, and it may indeed be a consequence of pesticides and glyphosate. Worker honey bees can forage for several miles if required to locate food. This fact also means that their circle of exposure is pretty wide. When a foraging honey bee gets exposed to pesticides like neonicotinoids, they can become disoriented and fail to return to the hive, which soon thereafter collapses. This family of pesticides causes neurological harm to insects but is not supposed to have any negative effect on mammals. There is some morbid curiosity as to how bees are supposed to know this fact. If our diligent pollinators like honey bees are feeding off corn protein bathed in neonicotinoids, trouble is not far behind. Add in a life-sucking mite and hotter temps, and honey bees are being bombarded from many directions.

The loss of honey bees in regions dominated by large monocrop farms shows how susceptible they are to chemically intensive farming. Big Ag farms do not produce a very diverse buffet for honey bees. Monarch butterflies navigate a thousand-mile journey but a quick drink from a pesticide-sprayed plant likely signals the end. Imidacloprid, the most widely used neonicotinoid, was brought to market by you guessed it: Bayer AG.

Even the EPA, who once upon a time administered legislation to protect the environment, is reviewing neonicotinoids as of January 2020. Here is the official EPA statement:

EPA is proposing:

- management measures to help keep pesticides on the intended target and reduce the amount used on crops associated with potential ecological risks;
- requiring the use of additional personal protective equipment to address potential occupational risks;
- restrictions on when pesticides can be applied to blooming crops in order to limit exposure to bees;
- language on the label that advises homeowners not to use neonicotinoid products; and
- cancelling spray uses of imidacloprid on residential turf under the Food Quality Protection Act (FQPA) due to health concerns.[26]

Now, mind you, the agency's language is tame and their recent efforts have been laggardly in removing harmful chemicals from the environment. I wouldn't count on neonicotinoids being heavily regulated any time soon. Don't wait on the EPA; stop spraying these pesticides on food. Stop harming pollinators like monarch butterflies, and stop the chemical onslaught on our flora.

Monarch Population Decline

A study in 2016 estimated that monarch butterfly population in the Americas had drastically declined by 95 percent over a twenty-year period. Annual numbers can vary due to harsh weather or a single devastating storm but the twenty-year trend was troubling. The wintering season of 2018—2019 was very successful for monarchs, demonstrating seasonal fluctuations in numbers.

We could fault them for being so reliant on a single plant or we could fault them for sailing so far to breed and escape winter but we can only fault ourselves for the drastic changes in their population. First, we are driving climate change to unprecedented levels; second, we are reducing their habitat. It shouldn't be a surprise to anyone that monarchs are endangered. Those two factors are why a third or more of all species are at risk of extinction.

As noted, the elimination of milkweed also plays a big role in their marked decline in numbers. Imagine (before cell phones) that you are on Interstate 15 leaving Las Vegas on the way to Los Angeles in the dead of summer and all of the usual gas stations are closed in California and you get stuck in Death Valley on an empty tank. OK, maybe that's not the best analogy. Imagine a beautiful butterfly that has marvelously evolved over millions of years to gain iconic status dying out in such rapid order. Habitat destruction like the elimination of feeding and breeding grounds, and intense weather patterns like the winter of 2002 that destroyed 75 percent of the migratory population, are making the species very vulnerable.[27]

If we can achieve some balance between corporate and organic farming, and habitat preservation, we can boost the monarchs' numbers. We must search for a means of detoxifying our agricultural production to heal heavily farmed land. In doing so, we can support biodiversity and enjoy one of nature's most unique creatures forevermore.

Homeland Farm

My uncle and aunt, Bob and Arlene Clark, still live on the family farm officially known as Homeland Farm in western New York. Bob, known as Butch to most, turned eighty last year, and he and Arlene are both still active and spry. He was raised on the farm, as was my mother and my aunt Connie, and I was fortunate to spend a fair amount of time there in my youth also (mostly getting in the way of my grandfather who tended to the farm for the majority of his ninety-six years). They planted a range of crops, milked dairy cows, and made a comfortable home in rural New York State. Mom would attest to the hard work it took to maintain the farm, which also required outside income, and she can boast about learning to drive a tractor by the time she was ten.

Homeland Farm is now my uncle's farm and he has kept it up as his father did. It's a bucolic spot that I still enjoy visiting. Uncle Butch's milkweed patch is thriving again. He recently told me that last year he saw more monarch butterflies than he had seen in years, which was news to my ears. My uncle's patch makes a difference; your patch can as well. Growing milkweed in Central Texas, a key stopover point, has even more significance. Monarchs arrive there from their winter haven in Mexico and lay eggs on milkweed to begin the generational migration all over again.

Texas is one hell of a land mass, so more milkweed portends better outcomes for monarchs. Craig Wilson, of the USDA Future Scientists Program, has been monitoring butterfly populations in Texas for years. As an expert on monarchs, he says that it is critical for gardeners in Texas and other states to plant milkweed in their backyard to sustain their population.[28] That's good enough for me. It may be a symbolic gesture here in Massachusetts but I enjoy planting milkweed and attracting monarchs every summer. I haven't seen any signs of eggs or pupae but they certainly hang around a lot, and it makes me feel good. It is this biophilic attraction I have to monarchs that drives me to speak on their behalf. We don't exactly get a menagerie of exotic creatures in our backyard so monarchs are a colorful addition. In 2020, the year of the virus, backyards have become an oasis for many of us. We warmly welcome you back, aerial migrants.

CHAPTER

One Green Deed   2015

*F*or the past year I've been writing a nature book about pre-
serving our Earth, the third big rock in our solar system. We
may be clever enough someday to colonize other rocks and
space walk our species into an interminable future. I seriously
hope so. But I am truly worried about this planet: the one cel-
ebrated on this day, Earth Day, in 1970. After decades of con-
suming books, lectures, and documentaries, I've accumulated
a good deal of knowledge that I felt compelled to share with
others, like-minded or not. And along this bumpy road, I met
some incredible environmentalists: scientists and nonscientists,
activists and nonactivists, writers and nonwriters. Now, forty-five
years after the first Earth Day, I'm halfway through my book. It
has been an extraordinary privilege to reach back out to some
of these environmental pioneers and ask them to reflect on the

same question: If you had one green deed you would like to see heeded, adopted, and passed on, what would it be and why?

Back in 1997 when I embarked on my quixotic expedition to explore the happenings of our planet and reinvent my career, I was so naive. Chasing environmental windmills is the proverbial act of naivete, many of us now know. It is a heavy task to convince others that we're not taking care of our planet. Today, the term global warming has become a joke if you're a nonbeliever, and a hot potato if you're of a different view. To hear non-scientists like me make the case for global climate change stuck in the oncoming traffic of fossil fuel-derived energy companies is akin to the Monroe Doctrine for Spanish conquistadors. It never mattered to me that I would face opposition to my approach toward how we must care for our planet and its bounty. I still am naive to a degree, better informed but strategically naive.

Well, I decided the problem wasn't the science, it was the relationship to the science. The general audience shouldn't be daunted by science and by no means should be afraid to participate in environmental conservation. Outside of astronauts, our generation will never have to space walk. Outside of the sun going supernova unexpectedly, our generation will survive. I am distraught about compromised survival though. Future generations will inhabit what we were delinquent in managing. That point never ceases to bother me, and that point should be obvious to us all. I mean, if you scroll down on any credible online news site, you'll see a report on nature indicating that yet another alarming milestone has been eclipsed.

I have hosted, participated in, and keynoted many environmental discussions since 1997, and I have spoken about sustainability on two continents. I founded a business that espoused sustainability as a fundamental principle. None of this would have been even remotely interesting to anyone except for the passionate environmentalists who I met along the way. The book,

One Green Deed Spawns Another, is the parallel story of the circuitous route I took to become an environmental consultant and writer, and the one that brought me in contact with amazing minds, each of whom share their vision for the one green deed they'd like to see universally adopted. Anyone would have been affected as I have been all these years in the wake of this knowledge, and to hear it from so wide a perspective as I have is worth sharing. Therefore, my one green deed is to connect, and, in earnest, one green deed will spawn another.

Writing One Green Deed

In 2015, I was deep into the interviews that would become twelve of fourteen chapters of *One Green Deed Spawns Another: Tales of Inspiration on the Quest for Sustainability*. The origin of the book was the outcropping of assignments for my online MBA in Sustainability from San Francisco Institute of Architecture. The written requirements of the degree forced me to reflect on the journey that I had been following for close to twenty years—a journey filled with experiences not that common to other environmentalists.

As a nonlinear thinker—you've surely discovered that by now—I chased the scent of anyone who genuinely inspired me. In a strange way that served me well. In one of the first environmental talks I gave, I spoke of broadening the green horizon. And in my mind that was just what I was doing. By the way, that brief talk in 2002 ended with the phrase "one green deed spawns another"—as did most others subsequently. Embracing sustainability or ecological balance truly means broadening the green horizon because ecologists and economists have different silos of thought, to name one pairing, but protecting Earth should have no conceptual boundary. It dawned on me that I could ask the same question of Native Americans and marine biologists as I could of architects and social activists. If we agree that preserving a healthy life experience is a common goal then we must band together. Their responses are what made the book so compelling for me to write.

I have heard from many readers and friends of *One Green Deed Spawns Another* who have impassioned concerns for the future. Some of the feedback was as moving to me as the interviews for the book. When a reader shares with me that my book compelled them to join an organization or make a lifestyle change, I proudly reflect on the original reason I chose to tell the story. What also never ceases to amaze me is that all who took the time to contact me represent every facet of society. And that is the focal point of why we must band together to navigate the future. Climate chaos or disruption or devastation affects us all. We in the United States may be spared the initial brunt of it as a large land mass with abundant resources but we won't be spared for long.

Speaking to people at various book readings in 2018 and 2019, I am also reminded that I am preaching to the choir. The collection of well-read and well-heeled environmentalists I've encountered is such a source of hope for me. I've needed that at times because I have also been confronted by let's just call them "naysayers," although we have subsequently learned that they have ulterior motives. Money is as toxic to the environment as if it were its own greenhouse gas. If fossil fuel companies can influence political will, then they are emitting additional dangerous gases.

When someone poses a recognizable climate-denial talking point as a question after one of my talks, I get enervated, to put it mildly. Again, I can forgive mistakes of judgment just as I can honest and legitimate refutations of the subject matter. As a nonscientist, I used to defer to anyone who had scientific background when speaking on the topic. Years later I decided against that approach, and it had nothing to do with errors of judgment. It takes a few knockdowns to get up and fight for the cause. I gained a lot from the many events I attended but the lessons for me are still coming.

Apathy

One of the most difficult lessons I learned from the research, interviews, and public events is that environmental activists face two colossal enemies: apathy and greed. The former is the harder of the two to understand. I have spoken on the importance of sustainability for decades. My articles on ecology and sustainability have been published in print and online publications. I have pleaded and posited at various events, trade shows, and conferences about adopting sustainable lifestyles and practices. I have interviewed climate scientists, biochemists, foresters, farmers, sustainable chefs, professors, and fashion designers at various industry events. I've toted around my favorite books on the environment to furniture expos, and I have even been shaken down at the Canadian border for carrying hemp fiber in my carry-on bag. I explained sea turtle gestation to an attendee at a furniture tradeshow in 2001. My talks have been translated for non-English-speaking audiences, but I still feel that I've hardly made a dent on people's attitudes despite all of this. It may have all to do with the fact that I'm not very good at what I do. But, was anyone really paying attention? Was anyone broadening their horizon? Does anyone care?

For the past eight years I've helped create sustainable hospitality discussions for NEWH (Network of Executive Women in Hospitality), a large hospitality networking organization I've served and supported for many years. The broad success and influence of NEWH includes providing scholarships for hundreds of hospitality students and launching the careers of many dedicated hospitality professionals.[1] My volunteer work for NEWH has evolved into developing talks with green themes related to the hospitality industry. In 2016, I invited Dr. Max Holmes, Senior Scientist of Woods Hole Research Center, to join me in New York City at a hospitality design trade show to talk directly about climate change. Max is now Deputy Director of Woodwell

Climate Research Center—as it is now known. In any case, it occurred to me that most people in the hospitality industry had never heard directly from a climate scientist of his repute. I had gotten to know Max through my association with his organization and had come to enjoy his company. Anyhow, he deftly presented the consequential topic of climate change through the perspective of a climate scientist to the assembled crowd of trade show attendees. Response was very good based on the feedback we both received. All in all, I was pleased. But did we make any impact? Did anyone attending the event contemplate changes in business or lifestyle practices in any way?

Max and I have met up a few times since and have stayed in touch. I emailed him in November 2020 about his opinion on whether the bigger problem is apathy or greed. I was curious of his perspective as someone who has dedicated his life and career to environmental science. His response follows:

> Apathy and greed certainly both play a role, but for the majority of Americans I think apathy wins out. It may be that they are worried about climate change, but they feel like they have other more immediate concerns so it gets put on the back burner. And though the impacts of climate change are already here today, what we're really worried about is a bit further down the road. How do we move those concerns to the top of peoples' lists, particularly in the midst of a global pandemic, a crazy election, etc.?

I recall another talk that I developed in 2011 for the NEWH Leadership Conference in Florida about the balance between developing and preserving Atlantic coastal communities. Florida is dotted with gorgeous coastal cities that attract a high number of tourists. With this development comes the real possibility of habitat reduction and encroachment. I invited marine scientists from University of Florida, Florida Fish and Wildlife Services, and

Florida State Parks to offer their expertise as well as to counsel developers on critical sustainable practices. This specific conference had concurrent sessions, as do most. When all was said and done, the session that ran at the same time as our presentation, which was about social media, attracted the bulk of the conference attendees. Our discussion was fascinating and, in my opinion, of far more long-term consequence but understandably less exciting to many attendees. I felt obliged to apologize to the esteemed panelists we brought in for the poor attendance but much to my surprise they were pleased. None had ever been to a hospitality industry conference that hosted a similar talk. Folks, we have a long way to go. *How can preserving the planet ever compete with marketing on Facebook or Twitter, said no one in 2050?*

Environmental apathy is a symptom of shortsightedness. If all we can conceive is how to see through the eyepiece of one generation, we may very well be on the course of destruction. Apathy, and its second cousin, obliviousness, are decisions that we make. A lack of curiousness about the world around us is a conscious decision. We don't all have to be Darwin to care. We don't all have to be Rachel Carson to care. But we have to be present, be engaged, and share our thoughts and wisdom with others. When we don't speak up, we are offering tacit support for the status quo. And that isn't serving any of us well at this critical moment in time. Before we have to fight to preserve our species on the only planet that sustains life in our solar system, we have to fight to care.

I understand people in desperate straits not addressing long-term issues like climate crisis. Though I might question the idea that climate crisis is only a long-term problem, the actions of people fighting to survive on a day-to-day basis can be easily understood. Others with the means to engage in climate-related actions but do not are harder to understand. We all are trying to take care of our children; pay our bills; be good parents, spouses, and neighbors; be responsible citizens; and improve our lives, in general. In doing so, how can we ignore that our behavior today is compromising the planet's health? We cannot be absolved of this responsibility. I know I'm not.

I do often think about how I can change my daily patterns to reduce my environmental footprint regardless of the scope of the problem. We have a small, energy-efficient home with thirteen solar panels on top, and most if not all of our interior lighting uses LED technology. Our home was recently weatherized and insulated with nontoxic cellulosic fill by a fellow environmental activist's company. We compost and recycle much of our waste, and ultimately, we try to use our consignment store for as many purchases as we can. We have paid off both of our older midsize vehicles (EV next time), and we volunteer and support local environmental groups and attend or lead green discussions. We rely on our city's farmer's market for local grown produce, eggs, and other food or home products, throughout the summer, and shop locally year-round for other goods. We use solar, battery, or electric power for lawn care equipment as well as for landscaping, and we plant native flowers, herbs, and trees without the use of pesticides or other toxic additives. Virtually all of our appliances are ENERGY STAR certified, and we have limited our water usage overall. I call my local, state, and national politicians when important pro-environment legislation is up for vote. I have shared ideas with our local mayor, who has proved to be a willing environment leader. I'm still contributing far too many tons of greenhouse gas emissions irrespective of all that. I fly to get across the United States—hey, my son lives in Denver, and even with offsets, I'm not walking there.

None of what we do is very revolutionary or even remarkably unique. I have friends who live carbon-neutral lives and are far more creative than I am. The point isn't to shame anyone into adopting sustainable lifestyles. That isn't what motivates people. We need to be inspired to change, and just as important, we need to be unafraid to lobby for pro-environment legislation. I reflected on this throughout the fourteen chapters of *One Green Deed Spawns Another*. Stories of people initiating change through innovation, sacrifice, research, commitment, care, and concern are worth sharing. What would move someone to dedicate their life to a sustainable enterprise or education is what I wanted to know and reveal.

Inspiration is an antidote to apathy and it comes in many forms. I found inspiration in the examples of selfless compassion by others. Ones who long ago determined to sustain the planet and its diversity. The same mission I embarked on in my thirties. Their stories and their impact on mine are the basis for the green deed I could pass on to others in the form of a book.

Most of the book feedback was overwhelmingly positive—notes of truly wanting to make a difference. My sister-in-law and her wife planted a wildflower patch in a neglected area near their home and named it the Green Deed Wildflower Field—I'm still touched by that. And we forget sometimes how important visualizing change can be.

Greed

Greed is a more direct and conscious threat to environmental preservation. If someone is profiting by degrading the planet, then they have a burden to bear. *Burden* is too soft a term. They have a sentence to serve to the children of the future. It was clear in 2015 when I was writing the book that companies were well aware of the potential for climate chaos by continuing to extract and burn fossil fuels. Not only did many of them know, they even shared the information for years through internal memos and studies.[2]

The science behind climate change wasn't the purview of just climatologists. Geoscientists were studying the potential for atmospheric warming all along. These representatives of Exxon, Texaco, and Shell quite accurately predicted global warming scenarios as long as forty years ago.[3-5] What angers many of us are the conscious decisions to hoodwink the public in the face of the very science they conducted. The average sixth grader could conclude that a rise of 10°C at the North and South Poles spells disaster.

Instead of sharing their findings, and rapidly transitioning away from fossil fuels, which should have been the ethical directive, they began to fund climate denial science. Not only did they target scientific data, they targeted climate scientists as well.

Dr. Michael Mann, atmospheric scientist and esteemed professor at Penn State University, has chronicled the strength and severity of the fossil fuel-funded opposition in his most recent book. He has vigorously defended his scientific research, including the well-debated hockey stick graph, which has been found to be not only accurate but an eminently helpful tool for us nonscientists. No one should have to face a well-funded opposition with ill intent, and Dr. Mann has experienced more than most. [6,7]

If we didn't think science could be manipulated before then, we sure know it now. Billions in dollars were invested to carefully place uncertainty among the facts of anthropogenic climate change. It wouldn't be unusual for the average layperson to accept these professional-looking documents as authoritative. These concocted articles and campaigns have become like a shell game with we as the dupes trying to get at the truth.

Sowing the seeds of uncertainty has proven to be very effective, as it turns out. Gallup polls covering the last two decades show that half of all Americans still didn't believe that climate change was a result of human activity, specifically, the burning of fossil fuels, as recently as 2010.[8] In light of all of the climate indicators revealing that each new decade is the warmest on record, this is stunning. Even fewer Americans believed that climate change would cause any threat to their lifestyles as recent as 2017, despite the fact that millions die each year as a result of climate-related issues.[9]

The World Health Organization lists climate change as one of the greatest health risks of the twenty-first century.[10] If you live in an impoverished nation, you hardly need to be polled. If you live on one of the islands of Tuvalu, you don't need additional climate data, you're simply looking to migrate to higher land. Disguising fake science as real science is a crime against humanity if you're profiting from it. And that is what these fossil fuel companies have done by sponsoring climate denial reports. But many of the large fossil fuel companies are facing a threat they haven't had to contend with in the past: environmental law. Fossil fuel companies denying their duplicity in the climate crisis are now being challenged in the courts.

Fossil fuel extraction and usage must be assessed for its consequential damage. A sustainable green economy places a price on pollution and environmental and social impact. We must reject the traditional economic model that allows for fossil fuel companies to tip the scales in their favor. Does anyone really believe that solar, wind, geothermal, and other sources of renewable energy cost that much more than oil, gas, and coal? Current carbon tax models show how the costs become very comparable when dirty energy polluters pay fifteen dollars a metric ton of greenhouse gas emissions.[11] This will be discussed in depth in chapter 9.

This quixotic journey isn't always futile or as satiric as Cervantes depicted it in the 1600s. We, the errant knights of the environment, find windmills worth chasing. But why? The joy for many of us is in the quest. When we lose passion, we retreat in a way. I faced down windmills writing *One Green Deed Spawns Another* with the same naivete. The tale of my path to environmentalism isn't nearly as adventurous as that of Don Quixote but it was no less impassioned. I think we all reach an epiphany, a lucid moment, a reason to charge ahead into the abyss. It doesn't matter what motivates one to take that step; what does matter is that we bring everyone along.

The spirit of collaboration is a powerful implement in the construction of any successful movement. It also renews our faith in others. I expressed this in a poem I wrote many years ago watching one of my son's soccer matches. And although it makes no specific mention of climate activism, it refers to inspiration and the importance of engagement. We all need to get up and fight for our future.

Lessons from the Soccer Match

The quest for sportsmanship is an interminable track
to grace—

an opening to see in others what we lack in ourselves.

We have come to be—in fear of exposure, emotion,
and weakness—

but truth is rescued in an instant from boys, whose
tears fall free,

arranged in a circle after losing the championship on
a penalty kick.

Hope, indeed.

Our inspiration has become artificial, digital, cast
from alien souls,

sidewinders of faith and friendship, we are unable to
rise from a glancing glance.

We lack courage when we don't extend ourselves,
seek others.

They are still whole, not yet compressed by life, fresh,

unable to accept defeat, prostrate on the turf.

Hope, indeed.

We worship heroes lacking credentials and color,
biding time

for a Walmart reply to a complex question of
conscience,

a retardation of responsibility, commitment, concern.

Nine-year-olds cannot know this, feel only the
moment,

and the steady arms of a guardian, parent or big
brother.

Hope, indeed.

Our age is sequentially marked by failings and
wisdom,

and our reflection is angrier than it is supposed to be,

catalyzed by shaky knees, decaying bank accounts,
and blood tests.

Yet today young men ripe with letdown, loyalty, and
grass stains,

rouse us armchair parents to get off the bench and
participate again.

Hope, indeed.

(2009)

8

CHAPTER

Dedicated to Peter C.H. Pritchard, 1943–2020.

Through Myrtle's Eyes   2016

Myrtle has survived well into her eighties—no one really knows for sure how old she is. We went to see her the other day, and as Earth Day 46 nears, it made me think about all that she has witnessed in her long life. Myrtle has been around since the Great Depression and she survived World War II. She may even have been on a beach for the 1969 lunar landing of Armstrong and Aldrin. Well . . . Myrtle hasn't actually seen any of these special events because she is officially a *Chelonia mydas*, commonly known as a green sea turtle, and lives at One Central Wharf, Boston, Massachusetts, 02210—the New England Aquarium. Myrtle is somewhat of a celebrity in marine research and conservation circles because it is widely believed that she is the oldest sea turtle in captivity. Sea turtles in the wild don't often live that long, so Myrtle has truly seen more than her kindred species. Sea turtles are reptiles that have been around since the dinosaurs, and they've witnessed much in their 100 million or so years. And Myrtle may know more than we realize as a marine reptile. One expert, Jean Beasley, an octogenarian herself, who founded the Karen Beasley Sea Turtle Rescue and Rehabilitation Center in 1997, said to me recently, "I think most animals are far more sentient than we give them credit for. Yes, they recognize voices. We've had turtles that would not eat from someone whose voice they didn't recognize."[1]

Myrtle started her mad dash on some southeastern Atlantic beach at a time when green sea turtles were abundant and dunes were still pristine. Seagrass and green algae dotted the shallow coastal waters, and marine travel was dominated by living, breathing creatures swaying with the Gulf Stream currents. By the time Myrtle came back to that same beach to begin motherhood, the ozone layer had thinned, the nuclear age had begun, coastal development was reshaping the beaches, moonlight was outshone by exterior spotlights and neon signage, and life on land was much more confusing. By that time, Myrtle had already beaten the odds of 1,000:1 and avoided the market for her meat, shell, oil, and skin. But, like all female sea turtles, she would never know if her eggs were allowed to hatch.

As Myrtle turned middle-aged, she competed for food with endless barbed fishing lines and massive nets scraping the ocean floor disrupting everything in their path including many of her chelonian friends who wouldn't live out the day. Her life was made even more confusing with a new type of jellyfish that she couldn't digest: clear plastic. And unexplained tumors were showing up on many of her closest friends as her ocean home became more polluted. Concerned marine scientists would establish the ESA, the Clean Water Act, Earth Day, and ban DDT, aerosols, and ocean dumping, and still her population continued to decline. Citizens of coastal communities would slowly begin to protect nesting sites, as Karen and Jean Beasley did on the Carolina Coast. They would educate others about the odds of one hatchling making it to Myrtle's age. And back then few sea turtles would get a second chance since rehabilitation hospitals were virtually nonexistent. All of the seven species would end up on the list of endangered species.

Myrtle in her eighties no longer faces the perils of today's oceans and she gets fed daily; in fact, she gets fed first because she hounds the aquarium divers otherwise. And nowadays she circles the glass enclosure carefree, rests without fear, and monitors the passersby unfazed. Likely she doesn't know that her former habitat is warming and rising or that her natural food source is dwindling. Possibly, she is unaware that coral reefs are dying worldwide and that plastic pellets are as common as sand on the ocean floor. Maybe she knows that her daily visitors represent a species that has reached seven billion and has reshaped the planet's landscape permanently. I'm sure Myrtle would accept freedom willingly but she has been forced into retirement like we all will be someday. In her sagacity as an elder sea surfer, Myrtle has a lot to say to us on the outside. And I can see it in her eyes as she floats along the glass one last time: just keep paddling on.

Sea Turtles

Sea turtles have always been one of the more unique reptiles on the planet. Their long life span, migratory journeys, and inherent global positioning skills make them one of the world's iconic species. Let's see you retrace your steps to the place you were hatched, er, born without a birth certificate or your father's home movies.

Additionally, leatherback sea turtles can dive at similar ocean depths as sperm whales and travel over thousands of miles in a calendar year. They are a globe-trotting seafarer, which also makes them more vulnerable than other species. The course of their lives will take them across many borders. One nation may protect their nesting sites while another may fail to protect the waters where they mate. And international waters are a virtual free-for-all. With the added impact of warming ocean temperatures, sea turtles will need universal protection.

The sea turtles that nest along the Atlantic Coast of the United States have received far better care than in decades past. Active nesting sites are marked and protected, and the numbers of loggerheads and greens have increased steadily. For loggerheads, 2019 was an exceptional year.[2] If we can protect nesting sites, it does give sea turtles a fighting chance. Not every state has a Jean Beasley but everyone who visits the Carolina or Georgia coast should be darn thankful that she's been rehabilitating sea turtles for twenty-five years. If you've witnessed a hatchling or a nesting female on the East Coast, it very well may be as a result of her efforts. Myrtle's home, the New England Aquarium, has also relied on Jean's sea turtle hospital for years as a result of mass sea turtle strandings on Cape Cod.[3] The numerous sea turtle conservation organizations, including Jean's sea turtle hospital and Peter Pritchard's Chelonian Research Institute, which also includes turtles and tortoises, have been critical in the effort to preserve species.[4]

As to sea turtles, a prehistoric species, their fate is our folly alone. We altered their existence in rapid order; none more than the Pacific leatherback sea turtle. To make matters worse, we did it blindly and carelessly (other than poaching of eggs). The swamp of ocean trash certainly can't seem like home to these ancient explorers now caught in the torrents of a plastics tsunami. And the miles of ghost nets will prove to be a persistent threat for years to come. Many years ago, I wrote the following poem.

Dermochelys coriacea:
Plight of the Leatherback Turtle

Only a tidal invitation from the moon

can induce the beach walk

of a fugitive dinosaur.

Like a rare excursion for an aging actress

a post-Cretaceous grandmother magically appears.

But she is a death-row inmate,

her nests hatch condos,

her teens are bycatch

for a $6.99 fried shrimp platter, and we

Homo sapiens are settling a Darwinian score.

Shedding tears she thrusts her carcass to the Pacific

a final curtain to a million-year-old act.

As the moon careens skyward off her unarmored shell

it is a distress signal

to us kings of a lonely kingdom.

(2002)

Plastics and Ocean Pollution

Sea turtles are what scientists call a bellwether species, an indicator of things to come. Because sea turtles rely on both land and water and have survived since the dinosaurs, a sudden decline in their numbers is significant. It spells trouble. In a virtual blink of an eye, sea turtles became endangered. What makes sea turtles a study case is that even though they face predators on both land and water, they really only have one predator to fear—the populous one on two legs with the big brains. Their decline coincides with the rapid growth in global population, the effects of the Industrial Revolution, and the über-commercialization of our coastal cities and fisheries. One could argue that poaching, recreational boating, and natural predators are contributing factors but not as an explanation for their steep decline in numbers overall.

One very troubling sign of our carelessness toward marine species, including sea turtles, is the plastic invasion of their habitat. Ocean plastics are now a serious threat to ocean wildlife. Some portions of our ocean waters are a vortex of manmade pollution often called marine garbage patches. These patches are as large as some nations.[5] Some estimates suggest that plastic will soon outnumber fish in our oceans.

When did it become OK for producers to pollute at will and pay no price? If you make something that doesn't naturally biodegrade and becomes toxic to living creatures, then you are an environmental criminal. Sorry, that is a fact. Who shares the burden of guilt when a whale washes up dead because of ingested plastic? I remember as a kid tagging along with one of my parents into an antique store, seeing signs that read: you break it, you own it. Despite my devious plan to break something to avoid being subjected to an antique store experience, I took the sign seriously. So, when did this warning lose its meaning? We are breaking the components of a healthy environment and its amazing chain of life, and not paying for it.

I sincerely hope that these creative advancements designed to remove ocean pollution succeed. Some seem promising but the problem is immense. We, as a cognitive species, have to contend with the fact that we found a way to permanently pollute 70 percent of our planet. Sea turtles are dying due to ocean plastic, studies show that they are eating plastic, a lot of it.[6] Weren't we warned? Didn't we learn from the Lorax and the Truffula Trees in 1971?[7] And sea turtles aren't fictional animals like Barb-a loots, either.[8] Recent indices show that close to half of all fish species and ocean mammals are endangered.[9]

Ocean plastics are a heinous crime against nature especially when you comprehend the enormity of the problem. A recent documentary from the television series *Frontline*, entitled "Plastic Wars" exposed the lies surrounding the plastics industry, their associations, and their environmental stewardship.[10] To think we continue to support an industry that has an absurdly low recycling rate, approximately 10 percent or less, is inconceivable.

It is also another case of creating a false solution to a problem of one's own making. Plastics recycling is almost a misnomer with such low recycling rates. Consider that we produce about thirty million tons of plastic on an annual basis, most with no downcycling option other than landfills or incineration.[11] Who would find this an attractive option other than the producers? If most consumers knew that their efforts to separate and wash plastic waste, place it in a barrel, and haul it out to the street is for naught, they might be pissed off. They should be.

The Society of the Plastics Industry developed the resin code, RIC, in 1988 to identify the types of plastic resins used for the product. We easily recognize the triangulated arrows at the bottom of a container—well I do after I put on my readers. Ostensibly this was done for ease of identification and recycling. No one bothered to determine if any of this profligate waste had any commercial value. It was an industry in search of a market and an easy excuse for plastics producers. They had no intention of being responsible for the life cycle of their products. Many recyclers found out they couldn't generate a successful business model, partially because virgin plastic was so cheap—artificially cheap, to be sure. A concerted effort was made by recyclers around the country, and although a few found avenues of success, the recycling rate has not substantially changed. That is until January 2018 when China refused to import any other nation's garbage. A great portion of what we clean and put in our recycling bins is landfilled or incinerated because it has no secondary market value and never did.[12] Only two of the RIC codes, 1 and 2, have even reasonable recycling rates.

Plastic packaging and products are becoming more important to the fossil fuel industry since cracking ethane into plastics here in the United States has increased. Ethane is one of several natural gas liquids, and it's the one used in ethylene production. The heating and splitting of these molecules is referred to as cracking.[13] It represents close to a third of profits. With oil trading downward, even reaching less than zero on April, 2o, 2020, cracking this crap is a new lifeblood for the fossil fuel industry. Few Americans want to see us ramp up our plastic production but this is a current trend with no immediate end in sight.

Even in a pandemic scare, the plastics industry will find a way to promote its products. In the fear of COVID-19-contaminated surfaces, the plastics industry has convinced the Department of Health and Human Services, and many governors, to suspend single-use plastic bans in a variety of states including mine, Massachusetts.[14] Can you imagine if the plastics industry transitioned to the production of personal protective equipment and testing kits back in early 2020 when they likely knew the need? We cannot get people here to wear masks in public and social distance effectively but we can free them from bringing reusable bags to the grocery stores. I feel safer knowing that we're liberating plastic bags in protest! The likelihood of getting COVID-19 from a reusable bag is very low but the plastics industry found a way to stoke fear. If you believe that the fossil fuel industry is lobbying for plastic bags over deep concern for life, just consider that they are ramping up production of one of the most lethal materials to ocean life ever conceived.

One of the more astonishing facts that I have managed to store in my pea brain is that nature serves as the best design for sustainable products. The whole concept of biomimicry is quite fascinating to me. According to Biomimicry 3.8, a leading organization providing biomimetic strategies and training to individuals and businesses, it is defined as learning from and then emulating nature's forms, processes, and ecosystems to create more sustainable designs.[15] Having befriended a biochemist from Biomimicry 3.8, I have been inspired and intrigued by nature's possibilities. Mark Dorfman, senior scientist at Biomimicry 3.8, has shared with me portions of his research, proving that nature is the wellspring of greener product solutions. From the reflectivity of polar bears to the strength of spider silk, from the perfect ventilation of termite mounds to swim suits designed like shark skin, there are many inspiring examples of biomimetic solutions to design challenges.[16] Having collaborated on a project with Mark, I can vouch for the innovative system they use to create more sustainable designs. Once when I was spouting off about the environmental hazards that chemistry was inflicting on nature, Mark glanced back at me and said, "But nature is chemistry."

We are fortunate to have chemists who make the distinction between nature-based chemistry and synthetic, lab-based chemistry. Mark recently penned an article on synthetic plastic packaging and ocean pollution through his unique perspective. Here is an excerpt:

> Flexible film packaging used by the food and consumer products industries to keep food, cosmetics, and other consumer goods fresh and intact, serve many of the same functions as the beetle exoskeleton. Industry achieves those functions by gluing together layers of disparate materials such as plastic, foil, and paper—each material serving a single function. While individually, these materials can be recycled, once bonded together, recycling is all but

impossible due to the difficulty of separating the thin layers, so they end up in landfills—or if carelessly discarded, perhaps eventually end up in the ocean.[17]

Plastic doesn't have to be fossil fuel based, there are countless biomimetic solutions lurking out there in nature. We have the duty to hold those companies most responsible for our polluted waters accountable for past and future waste. We know they never created closed-loop designs for their products. They felt comfortable shifting that responsibility to consumers.

I don't want to see my friends in Indonesia incinerating our single use packaging either, which is another form of environmental injustice. Our legacy shouldn't be a blanched landscape as a result of our toxic excesses, but the ocean's changing landscape is an indication that we've already begun to do so. What hell hath we wrought upon this ecosystem? No more sea horses mating with Q-tips and sea turtles swallowing plastic jellyfish, no more fish feeding on microbeads, and lastly, no more plastic tombs for hermit crabs.[18] We better invent more sustainable packaging options, for one, and come up with a plan to clean up our deep-sea messes along the way. Alas, we've probably waited too long. I just can't bear to hear about one more marine creature washing ashore gut-packed in plastic. Nature didn't design that.

Peter C. H. Pritchard

When I saw that Dr. Peter C. H. Pritchard passed away, I was returned to an earlier time in my life. I did not know Peter other than from an embarrassing experience in 1998 but I became a fan of his regardless. I don't know if anyone else suffers from insomnia as a result of past embarrassments, but I do.

In 1998, I named my little furnishings company Olive Designs; the "olive" of Olive Designs came from olive ridley turtles, not the edible fruit or vegetable, or whatever it is, version. I determined I wanted to provide funding from net sales to an organization committed to sea turtle preservation. Along the way I was introduced to some wonderful sea turtle researchers and conservationists.

Profiled in my first book, Jean Beasley is unlike anyone else. Her commitment to sea turtles inspired me to donate a percentage of Olive Designs' proceeds from sales to her sea turtle hospital in Topsail Beach, North Carolina. My business never got far off the ground so donations were small but Jean's sea turtle hospital had enormous support and made it despite a lot of unforeseen hurdles. Her center is more successful than ever, having rehabilitated more than seven hundred sea turtles.

When I launched Olive Designs, it was full of ecological principles and a lot of faith. The company brochure included a photo from Dr. Pritchard: an amazing photo of a female sea turtle emerging from the ocean to locate a nesting site and hatch her eggs. I contacted the Center for Marine Conservation to inquire about permission for the use of the photo. They graciously put me in contact with Peter. When we spoke by phone, it was clear to me that neither of us had any experience in determining a reasonable cost for the use of a photo. He asked me what I thought would be fair, and I asked the same of him. Within the call I also asked him if he was a sea turtle photographer, and he kind of chuckled and said that he was a sometime photographer. He ultimately threw out a modest figure, which I promptly accepted. Before I cut him a check, I looked up Peter C. H. Pritchard.

This is where the insomnia comes in. Peter must have found my question about his photography profession pretty humorous. For heaven's sake, Dr. Pritchard is considered by many as the father of turtle conservation. He studied under the legendary Dr. Archie Carr, who mentored many of the top marine biologists, and is equally renowned. Dr. Carr is credited with sounding the alarm of the decline in sea turtle populations. University of Florida will always be incubation central for sea turtle research because of Archie Carr.

Dr. Pritchard carried the torch for forty years upon his graduation from University of Florida, writing fourteen books including the *Encyclopedia of Turtles* in 1979.[19] He founded the Chelonian Research Institute in Oviedo, Florida, and was honored numerous times including by TIME magazine as Hero of the Year.[20] His conservation efforts are heralded for the continued survival of the kemp's ridley turtles of Mexico, and he has worked with several governments in turtle and tortoise preservation. He was one of the only conservationists to have witnessed and photographed Lonesome George, a male Pinta Island Tortoise, who was the last survivor of his subspecies. Three turtle species are named in honor of Dr. Pritchard. His esteemed career may have come to an end in 2020 but his legend lives on in the many students and followers that he influenced. Personally, I hoped he got my note of recognition included with the check—I'd sleep better if so.

Further, Dr. Pritchard's photo image was accompanied by a quote from Dr. Edward O. Wilson for my Olive Designs' brochure. Having long marveled at the research and written work of Dr. Wilson, I didn't make the same mistake as I did with Peter. Ed Wilson's official letter of permission has been framed for posterity ever since. I certainly had high standards for the back cover of a product brochure! If there ever were a famed biologist and Pulitzer Prize author to model your nature writing after, it would be E. O. Wilson. His pleas for preserving biodiversity are documented in science and delivered in poetry. And no one could have framed a quote better than Peter C. H. Pritchard.

Myrtle

Lastly, I am pleased to be able to report that Myrtle or the "diva" as Jean Beasley tagged her is still circling the tank at her adopted home. I reached out to New England Aquarium about her condition. Mike O'Neill, supervisor of the Giant Ocean Tank, Fishes

Department at New England Aquarium, informed me that she is doing fine. He went on to say:

> She continues to be the largest and likely most famous resident in the New England Aquarium, weighing in at about 550lbs (our Animal Care Team conducts a physical exam on her every 6 months). Her diet is comprised largely of vegetable matter including lettuce, Brussels sprouts, broccoli, bok choy, kale, and a rotating variety of other vegetables. We also provide her with some protein to round out her nutritional needs that includes squid and fish. It's also a great way to sneak her a multi-vitamin to make sure she's in tip-top shape. We estimate that about 60 million visitors have met her over the years and we hope she'll meet millions more in the coming decades and continue inspiring them to conserve our blue planet![21]

Keep on paddling indeed, Myrtle.

9

CHAPTER

Charge of the Climate Brigade   2017

On this forty-seventh anniversary of Earth Day, I am struck by how combative the topic of climate change has become. From my purview, our efforts seem so underwhelming in light of the recent shift in energy policy here in the United States. Having recently volunteered to help our local chapter of Citizens' Climate Lobby (CCL) a potent, bipartisan lobbying group committed to a carbon fee and dividend approach to reducing carbon emissions, I discovered how inadequately armed we all are to face off against the fossil fuel industry. Even with the strong bilateral support for the CCL carbon policy, it will be a bitter fight.

"Cannon to right of them, cannon to left of them, cannon behind them, volleyed and thundered: While horse and hero fell, into the mouth of Hell."[1]

Poet Alfred, Lord Tennyson certainly wasn't referring to a group of citizens marching in opposition to fossil fuel companies, and although the imminent danger is markedly different, the odds are similar. We aren't exactly light cavalry running headlong into heavy artillery but considering how slow the response has been to greening our energy industry, our opponent is equally as entrenched. Many of my environmentalist friends tell me that they feel so embattled today—curiously, facing Russian opposition again. We may not have to face off against cannon fire from all sides, but we are being opposed, to be sure. The political influence of the fossil fuel companies is at an all-time high. Having Rick Perry, Scott Pruitt, Rex Tillerson, and Ryan Zinke in cabinet posts is example enough. Their ongoing war on the science behind climate change is more than just self-serving, it has become a highly successful political ploy. It is no coincidence that these select individuals hail from Texas, Oklahoma, and Wyoming, considering their states' deep connections to the fossil fuel industry. Wyoming is the largest coal-producing state by a wide margin, for example.[2]

When an entire economy does not factor in pollution and the industrialization of natural resources, it is founded on a lie. The campaign to pit business against the environment is a tired old tactic of the fossil fuel industry. As if we have no choice but to burn fossil fuels to keep our citizens employed. The job growth suggests something quite different: Solar and wind energy jobs have outpaced the economy by 20 percent according to a study by the Environmental Defense Fund.[3] The muzzling of scientific data and environmental education as well as their spokespersons is just another way that the fossil fuel industry keeps their grip on our economy. We all know that the burning of their product is greatly responsible for the climate crisis we're in but we are divided by a political chasm heavily influenced by the Exxon Mobil's and Koch Industries. Eliminating fuel efficiency standards, dumping waste into our waterways, mining federal lands, expanding pipelines to transport oil and gas, approving toxic pesticide use, and delisting endangered species, to name a few, make no common sense to the majority of Americans. None of us voted for those measures to be enacted unless one had a direct connection to the fossil fuel industry. Yet, somehow, here we are watching this all-out assault on the environment, and the renewal of an economy enabled by our deep addiction to the extraction and burning of fossil fuels.

As I saddled up to go into battle by attending my recent CCL meeting, I discovered that Massachusetts legislators are mostly in agreement with the CCL legislative approach. And, overall, CCL's momentum is slowly gaining traction nationwide; as of now 36 members of Congress have joined the CCL Caucus. Many of them are from Florida—not a coincidence since they are already an early victim of climate change.[4] They've seen the data and they have seen firsthand the consequences. But the numbers are staggering when it comes to halting climate change to a two-degree Celsius level. We'll have to reach a near zero emissions scenario by midcentury to achieve it. That is a death knell to many of the world's most profitable companies—and they know that. They aren't about to tell us the truth about their impact on climate change nor should we expect it, but to know that as many people die annually from air pollution as from tobacco means that fossil fuel companies are as complicit in the destruction of human lives as tobacco companies have been for centuries. Then why are we letting them run our government? We never let tobacco companies have that much political power. If we don't act now, we'll be in the headlock of fossil fuel companies for decades to come until they are forced to show up in Congress with their tails between their legs. By then we may have waited too long. So, on this Earth Day, I'm going to march in reply—with good reason why.

Citizens' Climate Lobby

In early 2017. I reached out to CCL, a nationally represented, bipartisan collective formed to influence climate legislation. With chapters across the country, CCL appealed to me because they were placing their efforts squarely in the political arena. As a, let's say, older environmentalist, I was struck by the potential of an anti-environment presidency unlike anything anyone had ever witnessed. The handwriting was all over the wall. The thought of fossil fuel executives and their potent lobbying force shaping policy in a Trump presidency late in the game of containing irreversible climate change was frightening. The members of his cabinet bore this out. Some of the companies served by these cabinet members include CS Energy, Exxon Mobil, International Coal Group, Energy Transfer Partners, Kiel Brothers Oil Co., General Dynamics, Heritage Foundation, American Enterprise Institute. If you're an environmentalist, that information sends a chill down your spine. Virtually all facets of the fossil fuel industry were represented in the new administration. That didn't bode well for a pro-environment agenda, to put it mildly.

By the time I joined the local chapter of CCL, the efforts to deregulate environmental safeguards had begun. But it is cathartic to know that while your president is hell-bent on extracting every last drop of natural resources on federal lands, local and state representatives will listen and react to groups representing constituents. CCL does this particularly well all across the United States.

CCL was founded in 2007 by Marshall Saunders, a micro-financier and a grandson of Grover C. Thomsen, who was one of the founders of Big Red soft drinks. I've never had a Big Red soda but it is a version of a cream soda, I believe. I can't say I consume a lot, or any, soft drinks with high fructose corn syrup and red dye 40 but these are still very popular beverages. It may not be the healthiest of beverages but I'm going to give it a pass since Marshall's heroic efforts in founding CCL make good for it.

His story is worth retelling. In the mid 1990s, Marshall worked with poor communities, primarily in Mexico, to help them gain access to micro credit loans. His actions lifted thousands of the neediest out of poverty and starvation. By the time he created CCL, he easily could have retired into the sunset. But he saw the climate crisis as a similar matter of urgency. How did he arrive at that epiphany? According to his bio on the CCL site, Marshall Saunders was motivated after seeing the Al Gore documentary, *An Inconvenient Truth*.⬚ Most of us who watched the documentary felt an equal amount of angst and concern. The film brought global warming to a very public center stage and greatly affected people who hadn't paid heed to the warnings of scientists for decades.

One of those scientists, James Hansen, was an influencer of Saunders, and a primary source of information for Al Gore for the documentary. Although James Hansen wasn't as well known as Al Gore at the time, he had for decades run headlong into the fight with the fossil fuel industry and climate change deniers. Hansen was astonished to discover that he was being censored by the George W. Bush Administration; his scientific-based analysis was filtered through the lens of conservative politics and his own words rewritten to downplay the dire consequences of climate change.[6] This was a particularly disturbing new development, one meant for all top climate scientists. Global warming had become viewed as a partisan political issue versus a matter of science.

One could conclude that Al Gore's political ambitions were diminished because of the Clinton Administration and surely doomed as a result of some Florida election incompetence and the U.S. Supreme Court. But if you look at his years in politics, he had always staked his career on the environment. Instead of muffling scientists like Hansen and Roger Revelle, Gore gave them a stage and a microphone.[7-9] It's logical that his postpolitical career would include environmental preservation efforts. Since his failed effort at the presidency, Gore has won an Academy Award and a Nobel Prize and has founded The Climate Reality Project. I can't say that he ever gave up on his mission to curb carbon emissions, preserve biodiversity, and protect the planet, despite losing out on the top job.

Nor did James Hansen or Marshall Saunders give up. Hansen is considered the dean of climate science, and his scholarly articles have shaped many of the arguments for transitioning off fossil fuels, including the creation of the 350 parts per million of CO_2 threshold. The fossil fuel divestment group, 350.org, cofounded by Bill McKibben, refers to James Hansen as its scientific source of origin.[10] By the time Marshall Saunders had crystallized his actions into CCL, he had already dedicated years to social justice through his microfinancing business. Surely by then he had made the connection between environmental justice and social justice, as has been discussed in earlier chapters.

CCL has an interesting approach to attacking the climate crisis. Knowing that the United States provides billions of dollars in subsidies to fossil fuel production, to the tune of approximately twenty billion dollars, how do we even out the playing field of fossil fuel-based energy versus renewable energy? That figure does not even include hidden costs like pollution and health costs, manpower costs of protecting foreign sources of oil, and political lobbying costs, to name a few. According to Forbes, hardly an environmental nongovernmental organization, global subsidies for fossil fuels is in the seven hundred billion range.[11] Again, this figure doesn't include the hidden costs. If we include climate change-driven weather catastrophes, which I do, subsidies are well into the trillions of dollars. This is a truly alarming figure. We are using taxpayer dollars to offset the true costs of an energy industry dominated by fossil fuels. We rely on fossil fuels for more than 70 percent of our energy usage at a time when all climate experts say we need to transition from fossil fuels as fast as we can.

I think back to when I studied the book *Natural Capitalism* for my MBA. It was, and still is, one of the finest books ever written about how and why we should reshape our energy economy.[12] For one, natural resources aren't free. A sustainable economic model would properly factor in the costs of extraction and pollution, and those hidden costs I referred to earlier. Subsidies aren't a good solution to achieving sustainability. Factoring in true costs to the environment are, however. And that is where CCL comes in.

As much lip service as the dirty energy industry has paid to renewable energy, they've done little to boost it. A conversion to renewable energy isn't in their financial interests. It isn't in their shareholder's either. It is easier to lobby, mislead, and maintain the status quo. But, imagine if these hidden costs were levied against them. Consider that the Deepwater Horizon disaster cost BP fifty billion dollars, and almost bankrupted the company.[13] Maybe BP should have thought beyond petroleum long before 2010. In any event, a sustainable economy is a threat to all fossil fuel producers reliant on subsidies to control the market. A carbon pollution fee isn't the same as a tax if it takes in consideration all hidden costs of oil, gas, and coal subsidies. CCL believes that the best means of achieving an economic and ecological balance is a carbon pollution fee on fossil fuel companies. We call it a pollution tax.

As I have stated previously, no society, nation, company, individual that profits from environmental degradation should be free from accountability. And they damn sure shouldn't be getting our taxpayer dollars to do so. CCL goes further, however. The approach is to levy a pollution tax on fossil fuel companies by the metric ton of greenhouse gas emissions ($15 per metric ton is the CCL model) and redistribute the money to citizens or to a combination of both citizens and green energy policies. One of the best examples of environmental justice is to ensure that the hardest hit by climate change receive the benefits of policies like a carbon pollution tax.

Marshall Saunders had a long history of working with poor and disenfranchised communities so it makes sense that a hallmark of CCL's carbon fee and dividend policy is environmental justice. Dividends or rebates would disproportionately benefit citizens with smaller energy requirements.[14] A rebate would exceed any increases in energy costs for most citizens, studies by CCL indicate that about two-thirds of all Americans would make money or break even from a dividend.[15] The citizens with the least financial means are factored into the CCL equation. And remember that the people least responsible for our global climate crisis are the ones catching the brunt of it.

We in America are still emitting more than any other large nation per capita. Environmental justice means that we do have a higher burden than other nations to mitigate the circumstances of climate change. Instead of laying blame on China, which now is the biggest producer of CO_2 emissions, we should be working with them and all other nations to take on this crisis. For damn sure, climate change doesn't pay attention to borders or walls. One planet, one outcome for all.

Politics and Environment

I haven't always been willing to collaborate with organizations when it comes to the environment. I've seen organizations misappropriate funds or goodwill, and I've seen too many charismatic leaders succumb to power and profit. Many nongovernmental organizations cannot cross the political aisle. And that isn't always their fault. I accept the criticism that I haven't been as active as I should have been. I am a slow burner—I take a long time to warm to scenarios and people. Many of us need a catalyst to leap into action; mostly I have crawled. To President Trump's credit, he didn't usher in a happy middle ground for me and most of my peers nor did he for climate deniers. I have learned since that his surprise presidency was that catalyst many needed.

It was in that spirit I contacted the CCL organization to offer my support. That turned out to be one of the better moves that I have made. By good fortune a CCL chapter had formed on the North Shore of Massachusetts, and after one meeting I became a supporter and member. The volunteers of our CCL chapter were a collaboration of caring, smart, passionate, and talented individuals from every walk of life. Maryann and I developed such deep respect for many in our group, but maybe none more than for Rob Bonney and Jim Mulloy. Rob has led the local chapter of CCL for five years. Jim, his partner for twenty-nine years, has had leadership roles in CCL as well as 350Mass (nonprofit, grassroots organization dedicated to a better future without fossil fuels), and he is a resolute environmental activist. Their story is pretty unique and captivating. I reached out to them recently to learn more about their commitment to CCL, their motivation, and their incredible resolve. I wanted to share the discussion:

> **David:** What I was trying to ultimately get to, because some of the other chapters address this, is about the impetus for someone committing to a cause or whatever. We don't like to use the word *sacrifice*. We think that's too intrusive, but the fact is a lot of us sacrifice and for what and why do we sacrifice? For one, how did you get interested in CCL, and what drew you to that organization initially?

> **Rob:** So, well for CCL, it was generally more about changing our life trajectory. We'd both been trying to be considerate of our impact on the earth since *An Inconvenient Truth*, and certainly even before that.[16] By the way, it was also about the way we were raised—I was raised to not be wasteful. But I did indulge, I did a lot of driving, you know, some that wasn't necessary. *An Inconvenient Truth* was part one, where I realized that I was rationalizing some of my behavior, or at least that I wasn't considering the impact of my behavior in some ways. I started to think, is this really worth the pollution? And when Bill McKibben's article in *Rolling Stone* magazine, "Global Warming's Terrifying New Math," made its way to me, honestly, I was dumbfounded. I found it incomprehensible that we could have come this far, and that it was kept under wraps.

It kind of sent me into a tailspin emotionally because it was really, as the title said, a terrifying scenario, that it included the loss of civilization entirely.[17] So, two things most troubled me about that. One was the suffering that people were going to experience, myself, my loved ones, and humanity in general, and then the other was that eons of human accomplishment could be lost.

You know I read a lot of Shakespeare and I'm into literature and I just thought that it could ultimately be irretrievable. I couldn't really conceive of losing some of the most amazing creations of humanity, as well as nature, animals, and plants, etc. All of this could be lost, and after I got myself into a nice deep hole, somebody suggested that action was probably going to be most helpful to me. And then I started my journey.

Jim: So, yes, that was 2012 or 2013, more or less, but there are many different influences that go into decisions. And some of them are magnanimous like serving humanity. But some of them are very, very selfish. I think that is one thing people don't actually consider when they look at our story. Speaking for myself, there are a lot of selfish reasons for doing what I have done. I didn't want to spend twenty-five years working on research and have it come to nothing. So that's a very selfish thing.

David: Can you define your field?

Jim: Yes, my field is biomedical research. I'm a leukemia researcher, and I have a lab at Cincinnati Children's Hospital. I could have a lot of impact and do stuff that's very useful, but it's me not wanting to have to carry it into an uncertain future. That's very selfish. And probably most people in the United States don't realize that independent principal investigators have to support their own lab and their own salary with money that they raise, and that it's sort of a fundraising job that you get into more than anything else. Most of the money comes from government grants and the payline [the success rate for a grant] when I left in 2016, was at 8 percent. And now, in 2020, it is still at 8 percent, which means that 92 percent of grants get rejected. This is not a good feeling. And your life, well, income, self-worth, career, depends on your success. And, you know, when people's lives depend on something, things get worse and even nasty. There are a number of things along those lines that played into this decision for me.

My field is quite broken now. The systems in place to ensure fairness in publishing and granting, the two most critical activities in my field, have become highly politicized due to competition for funds. And it's such an unhappy space to be working in, and that makes it easier to move on. As money becomes tighter, the knives come out and it's really unpleasant.

Anyhow, there are multiple reasons but importantly we also had the means. We had the ability to do this. I guess some people would do it and some wouldn't. We did and we have a relatively good life. So yeah, we sacrifice, but there's a whole lot of selfishness in there on my part. I don't speak for you, Rob.

David: It was interesting that you equate your sacrifice to some level of selfishness. I wouldn't see it that way. But it's interesting that you position it that way because I think volunteerism is much about sacrifice. Well, as we all are trying to make a living, the more that you volunteer, the more that you sacrifice, and you guys volunteer a heck of a lot. Let me ask you both about what climate change is doing on a global scale because the poorest communities around the world seem to be bearing the initial impacts of climate change. Why is it so uneven in terms of how climate change impacts people of wealth and people of poverty?

Jim: I think it's all got to do with power and money—just about every single issue you can bring up. This is the underlying guiding force, and those who have it are going to make sure they get taken care of and are not negatively impacted. To the degree that others have no means and nowhere to go, they are out of options. I mean, this, this vast inequality in society is driving every single problem, in my opinion.

Rob: Yeah, I mean, not my area of expertise but I would say, following Jim's explanation, people with less money have to live in places that are more likely to be on a highway or next to a factory or near a toxic location. They don't have the means to change that nor are their voices really being heard because they've been intentionally disempowered.

David: I wanted to ask because you guys understand this so directly, and some people do not, that to enact change we actually have to work with legislators and engage people with political power. So, as you know, a carbon pollution tax is by far the single most important legislative task for CCL. How did you guys grasp that so early on, and how you've been able to get in that fight? What have been the key issues you've had to overcome? I guess it would be easier if I just asked one question instead of three wrapped into one, huh?

Rob: We were self-taught. I have followed the historical example that it takes years of building support, and it's slow. And even though climate change is extremely urgent, you have to build your team, build your army, whatever, one person at a time, and not give up. Keep believing that what you're doing is what it takes to succeed, even if it's frustratingly slow to start with, educating people that they actually need to get involved in the legislative world. Because, that is what has to happen. And that's hard. That was new to me and to a lot of others; then continuing to hope there's momentum with the number of people you reach, and that they're reaching out to other people. Without political will, nothing's going to happen

David: How have you found success in dealing with legislators, knowing that you've got to get in the political arena to produce any results?

Jim: The approach CCL takes is a little bit unique. As is 350Mass, which I've been involved with now for four years. Both Rob and I participate in 350Mass. We've been going through campaign selection that involves the theory of change and deciding what you think that represents. What's going to make the difference? Where do you see change coming from in society? We're in this one training that concentrates on nine different buckets, nine theories of change proposing the different ways we might effect change in the world.

CCL is approaching up to 535 targets, the members of Congress, we're not trying to convince the whole public, the 330 million people in the country. I don't need to bother with a whole bunch of dummies who just aren't and won't be listening. I think, really, we talk about political will and moving people, but CCL's target is really a member of Congress and getting enough people to influence them. And I think that movement has to be broader, it has to include people in the streets. I lean more toward that theory of change, people in the streets effecting change. Ultimately everything is policy, it does come back to policy and has got to be a policy solution. Is it carbon pollution pricing, which the experts say is the fastest way to achieve our goals? But that doesn't necessarily mean that it will be the way we're going to do it or that it's even the best way to do it but it is the fastest way to do it. And we do need speed right now.

I'm not wedded to one policy. I also understand the challenges are much bigger than climate change. For one, we are in the sixth mass extinction, and this one is human caused. And climate change is only reason number three for that mass extinction, probably not even the prime reason. So even if we take care of climate change, we've still got big issues to address. It's so much bigger than I imagined back in 2012. It's so much more daunting. But it's still doable, and I will keep pushing, just not necessarily the CCL way exclusively. Though it did get me into the fight initially. There's a lot of bad shit that we have set up that we've got to change, right?

David: We've got this infrastructure based on fossil fuels that we subsidize in the billions. Looking at renewable energy as the potential force that will ultimately begin to offset it, without mechanisms like a carbon tax, we're not going to get there fast enough. Tell me why you think this approach by CCL is going to help balance out our energy usage going forward to truly combat climate change?

Rob: Well, here in the United States, we want millions and millions of people to change their behavior and there really isn't a more efficient, more direct way to motivate people, than through the costs they experience. We need enough people to change their behavior, enough people to force the market to gain momentum and to complete the transition to clean energy. Our economy is like a huge ship that we are trying to turn so you need something powerful to turn it fairly quickly like the power of an unavoidable price signal that everybody sees.

Jim: There's too much power connected to money, and we're not going to get anywhere until we fix this. If we don't, they can block us at every stop.

Rob: The good news is that CCL is approaching 200,000 members. It started with twenty. We have representatives and chapters in every congressional district in the United

States. And we've worked with some elected officials for a number of years, moving them closer to voting for a carbon pollution price.

David: You guys really did leave other industries ultimately to commit to this, not just CCL, but commit to being environmentally active. What was that like to just say, I'm done?

Jim: Uh, yeah, well I already sort of mentioned that there were semi-selfish reasons that were motivating me. But, we're all smart guys, we all have accomplishments so I really didn't want to get to the end and look back and say why? Say why didn't I see twenty-five years ago that I should have made a move? It's figuring out that impact, and what feels like a good space for someone to make that impact.

Rob: Giving up our careers was scary but the alternative was also scary. I picked the scary thing that felt more personally rewarding and the one that brought me some peace from knowing I was doing the right thing.

David: And you both certainly are.

Placing mechanisms into a system that has not appraised natural resources won't be easy. But Jim Mulloy is right, in that money wields too much power over our institutions and without bold structural measures, we will not meet the challenge. One of the most egregious claims of the fossil fuel industry is that what is good for the environment is bad for the economy. To be clear, they are suggesting that any regulations on our traditional economic model hampers business and growth. Of course, that is the same economy doling out twenty billion dollars in subsidies to prop up an industry that hasn't been accountable for the long-term ecological damage it has caused. Today, renewable energy jobs are eclipsing fossil fuel industry jobs without, in many cases, an equivalent financial boost from subsidies. If the hidden costs of pollution are factored into the economic equation, it isn't even close.

I often hear about the workers being left behind by the shift toward renewable energy including coal and oil workers being affected by the Keystone Pipeline or another fossil fuel-based business closure. Folks, we can't have it both ways. We lost two generations of Mahood men almost exclusively to lung cancer; should I feel sorry for tobacco companies? The top climate scientists around the globe all state unequivocally that we need to stop burning fossil fuels if we are serious about halting catastrophic climate change. It was the first recommendation of Dr. George M. Woodwell when I interviewed him in 2014, and it is of all others I've reached out to since.[18] We can't maintain the old economy if it cannot control climate change.

I also believe in job training and transitioning, and workers in the fossil fuel industry must not be forsaken in the shift to a greener economy. Human resources will prove to be just as critically important to renewables. The fact that jobs in solar, wind, geothermal, and other renewable energy sources are on the rise in a depressed economy in 2020 is a positive trend.[19]

We discovered, among the tragedies of COVID-19, that locking down society and reducing consumption, while severe and unanticipated, had a profound effect on the environment. What lessons can we learn from this unintended consequence of a pandemic? If it takes strong measures to keep global temperatures from rising more than 2°C, which ones will we adopt from the model of 2020, if any? To me the lesson from the economic shutdown is that we now know what it looks like to halt emissions to a more sustainable level. Closer analysis would reveal that there are many factors at play to achieve the reduction of emissions of 2020 without the draconian measure of an economic cessation. We must pursue this challenge like it is, a global atmospheric pandemic.

And, by the way, climate change denial is a lost cause. Science could care less about public opinion. You can ignore and mock science but there is no herd immunity that will bring Earth's temperature back to normal. There is no climate normal anymore.

Yet, the world's nations can and must confront this phenomenon regardless. Rejoining an international accord is a necessary step but it's singularly not adequate. Societies, countries, governments, corporations, nongovernmental organizations, cities and municipalities, groups and individuals have to step up and take their part in preserving the environment, arresting climate change, and protecting biodiversity. Converting to renewable energy and putting a price on carbon pollution is but one of these measures.

Will we have the political courage to proceed? If we truly care about a sustainable future, we have to end the debate about transitioning off fossil fuels. Climate news is coming hard and coming fast. Experts predict that 2020 will be the warmest year on record.[20] The warmest years on record are now almost all falling in chronological order, and that is truly alarming. Typically, an El Niño year would skew this pattern but 2020 has now made it clear that anthropogenic climate influence is the most dominant factor in earth's rising temperature. There aren't two roads out of this crisis.

Maybe Marshall Saunders got tired of inaction and the influence of the massive fossil fuel industry on politics. He had the vision at the close of his career and ultimately his life to found CCL to take political action to transition away from fossil fuels through economic measures. I think it is genius. A carbon pollution tax is the closest thing we may find to properly assess the costs of our energy economy. No one representing us in Congress should be beholden to the fossil fuel industry, or any specific industry for that matter.

We at CCL are trying our best to see it doesn't happen. We can embrace green energy and transition away from fossil fuels. If you haven't sent that message to your local, state, and national representatives, do so as soon as possible. Perhaps you can join your local CCL chapter and get onboard. So, buckle your belt, sheath your sword, and mount up. The artillery of denial, misinformation and inaction lies in front, but charge ahead bravely anyway.

10
CHAPTER

Reluctant Sacrifice   2018

I recently attended the East Coast premiere of a documentary
entitled, The Reluctant Radical. It is a story of one man's
efforts to make people aware of the dangers of climate change
and the consequences of climate chaos.[1] It chronicled Ken
Ward's many years of imploring others to take action and to
better understand the dangers ahead. He ultimately became
so frustrated that he began a civil disobedience campaign to
garner interest and press. This is where the story gets deep.
Why would a well-educated, law-abiding, divorced father of a
teenager risk his personal freedom on potentially dangerous
publicity stunts to point out the perils of climate change? How
does someone arrive at that point in life? Nothing he said in

the various interviews were hyperbolic. His research is similar to mine. This film, wonderfully produced and directed by Lindsey Grayzel and coproduced by Deia Schlosberg, who risked their own freedom filming some of these stunts, is a startling reminder that it ain't easy to get others activated.

Here we are forty-eight years removed from the first Earth Day with more urgency than ever. From plastic garbage soups the size of entire nations floating in the Pacific to the critical endangerment of half of all other species to the rationing of water in South Africa, the risks are heightened. It is profoundly sad to me that we have to revert to illegal acts to make people take action in support of protecting future generations. As I write in my book *One Green Deed Spawns Another*, "I've learned that we're not all primed to destroy our brothers and sisters but in fact are programmed to nurture each other. Yet, we're capable of so much more."[2] So why did Ken Ward make the assessment that his freedom and possibly his life was worth sacrificing for this issue of climate change?

Ken Ward came to the same conclusion of facts that many of us have come to: our sacrifice for the planet's well-being is not close to the level needed to prevent the consequences of climate change—some that we are already seeing. It is particularly hard for many of us in the hospitality industry to reconcile this fact. Our industry is driven by comfort and luxury and not having to sacrifice what many of us do in our own homes. When we recreate and vacation, it is to seek some relaxation and quietude from our hectic work lives. I mean, I don't want to wash my sheets and dishes either. I don't want to shiver all winter because we won't set the thermostat above 67°F or haul our compost bin out every Tuesday morning (OK, Maryann does that) or look at the unsightly meters from our solar panels or haul around the ugly orange cord attached to my lawnmower. It is a minimal effort of sacrifice on my part and not commensurate with what has to happen universally.

So how do we create buildings and destinations that balance a minimal environmental footprint without sacrificing comfort and luxury? I've met some brilliant and talented people in the hospitality industry, and I think it's time to put our individual talents and expertise together and find out how. Consider what service or product or contribution you make to our industry and then consider, how would it achieve a zero-impact level? Obviously, that is not possible but how do we get substantially closer to it? This question has to become the driving message behind all of our products and services in the industry. How many of you would be comfortable in a Cape Town hotel today? Well, prepare for it.[3]

One poignant scene from the documentary was when Ken Ward's latest trial for sabotage for shutting down a pipeline in Washington ended in a hung jury. His son, who had to that point been pretty stoic and matter-of-fact about it all, gave his dad a very emotional hug. Many of us are not willing to get arrested or put our freedom on the line. I have friends who are, but I'm not. To me, we should never have gotten to this point in the first place. If we don't start sacrificing soon, we'll have missed the purpose of the first Earth Day march. They marched so maybe, just maybe, we wouldn't have to forty-eight years later. One thing that is completely intergenerational is that we all want a better life for our children and theirs. When should we start?

Ken Ward and Activism

I was fortunate to meet the film producer and director, Lindsey Grayzel, after a screening at the Salem Film Festival in Massachusetts, managed by my environmentalist friend, Stan Franzeen.[4] She recounted to me the harrowing experience of documenting an act of civil disobedience. If Ken was trespassing, then Lindsey was at well. She took great risk to tell this story in such a compelling way. I corresponded with her a few times, including sending her this Earth Day essay. I shared Lindsey's documentary info with a number of climate champions, and have since watched it again. It had the same impact on me the second and third time I viewed it. I am pleased to report in 2020 that all of the valve turners from the documentary are free citizens. A legal argument has been used successfully to defend their actions, essentially a necessity defense for climate.

Because the burning of fossil fuels contributes to the decline in health of humanity, there is a moral call for action in the form of civil disobedience. And as expected, Ken Ward is pushing the envelope again. In April 2019, he was one of five activists arrested for illegally trespassing onto private lands of a tar sands oil distributor called Zenith Energy Management.[5] Their illegal act, which was orchestrated by Extinction Rebellion, was to plant a garden at a rail terminal to block the expansion of oil flow into Portland. This act of civil disobedience was one of a growing number of incidents around the globe. Climate activism and civil disobedience are becoming more and more commonplace, and the lines between the two are rapidly intersecting.

So far, I've only participated in protests that have sought and received permits. I've witnessed other events that have not. Some of the most dramatic and distinct examples have been led by Extinction Rebellion, and it is no surprise that Ken Ward is among them. His past involvement with another notable organization, Greenpeace, speaks to his commitment to the cause. Most of us can recall daring Greenpeace demonstrations over their fifty-year history. One example was blocking Japanese whaling ships that had been disguising their activities in the name of cetacean research in 2006.[6] Commercial greed is typically the driving force behind the blatant exploitation of natural resources. We don't need Ken Ward to get arrested again to teach us that. They can hose us down, they can rubber bullet us, but their time is coming. The courts aren't going to prop them up forever, irrespective of the current role of our Supreme Court as a political backstop for the

Trump administration. I'm moved by those willing to sacrifice their freedom to protest the degradation of our environment. Far from facing down authorities and possible arrests, my protests mostly fill up pages and signs. You won't likely see me on railroad tracks or scaling a redwood but I'm just as pissed. How you choose to protest doesn't change the premise that we're in full agreement that our children deserve a healthy planet.

Legal Battles

Even before the mass protests in 2020 for Black Lives Matter and for racial justice, which I will discuss later in this chapter, there was a push in some states to seek legislation to limit peaceful protests. And these were not spearheaded by the Department of Justice and the second little man with the Napoleonic complex who led that department in the Trump administration. States took this on as a means of stifling opposition. I'm not sure why these specific states have chosen to administer greater penalties for protesters but it sure seems to have oil and gas written all over it. These pipeline states don't take kindly to people protesting their dirty products.

North Dakota even proposed a bill to allow drivers to "accidentally" run over protesters in the road. No, seriously, this was an actual piece of legislation.[7] It did not pass thankfully but let's not forget where the Dakota Access pipeline is on the map. Lawmakers came to the insane conclusion that if we can't keep Indigenous Peoples from protesting, we can at least propose to run them over. You'd think by now that we'd have learned that Native Americans fight hard for their ancestral lands.

A similar law was actually passed in North Carolina, albeit with greater restrictions. It puts the onus on a jury to decide whether there was any malicious intent to hit a protester or not. I wouldn't want to have jury duty that day.

Tennessee just added punitive measures to an existing protest law that would make it a felony for protesters to camp on state lands. In the guise of safety for state employees, police can arrest peaceful protesters who choose to camp out on state property land and take away their voting rights.[8] Voting is a sacred right; voting is the most important measure a citizen can take to enact political change. Some states, like Tennessee, want to squelch voter turnout to tip the scales toward minority rule. That doesn't usually end well, Governor Lee.

Cape Town Drought

In spring 2018, officials from Cape Town, South Africa, made the stunning declaration that the city of three million people had nearly exhausted its water supply due to years of drought conditions.[9] Many of us saw this as a clarion call for the world. Fresh water shortages will slowly starve regions stricken with drought. Agriculture remains the single largest recipient of fresh water, and as water tables and sources dwindle, communities become unsustainable. Many of the front lines of this catastrophe are in the Middle East and Africa. The effects of such droughts place millions of people and species in jeopardy. Mass migrations in that region aren't far off, and we've begun to experience migrations on every continent. Drought and desertification are factors that we will contend with across the globe. In Cape Town's case, water restrictions and timely rainfall averted a crisis. Populous cities of the future may not be so lucky.

Sacrifice

Sacrifice is an emotive word. It takes many shapes. No, I haven't shut off pipelines, and I haven't been physically confronted at protest marches or at my book talks or heaven forbid, a poetry reading.

Sacrifice, in my opinion, can be the means of further understanding climate change and mitigation. No one can confuse Ken's message and the other valve turners. We should all be adopting voluntary measures in addition to applying corporate and political pressure—it isn't gonna be voluntary very long anyhow. That hasn't always been the most popular suggestion of mine over the years. Yet, most of us didn't hesitate to shelter inside in the face of a novel coronavirus in 2020. Why don't we take climate change as seriously as COVID-19? If the planet has a virus called climate change that is killing species in the millions, and has only one cure, why aren't we marshaling the same forces to end it? Climate change attacks the respiratory system too, in the form of air pollution. One can make a connection between COVID-19 and human encroachment and loss of habitat. If we find that the origin of COVID-19 was a viral migration originating in live animals, one possibly endangered and illegally trafficked, then there is even more reason to make sacrifices in how we treat our planet and its inhabitants. There is no excuse for exploiting and abusing other species for our consumption. Our factories of the future should spit out solar panels not swine.

Motive

In 1998, I drafted an environmental statement for my small furniture manufacturing business, Olive Designs. In subtle terms, I established my motive for environmental activism. Others, like Ken Ward, were already actively following their passions. Ken was Deputy Director of Greenpeace USA at that point in his career. I came to the same conclusion as Ken and others that we weren't taking the looming environmental crisis with the seriousness that it required. Twenty years later, we still aren't. My goal was to stimulate conversation and share my sentiments through the products and principles of Olive Designs. The company and its products were a measure of my concern: pioneering more sustainable materials and designs, taking responsibility for product life cycle, and urging other companies to follow similar practices. The environmental statement reflected this. Here is an excerpt:

> We, at Olive Designs, envision a time when the effect of the Industrial Revolution on the environment can be halted. Sadly, it cannot be reversed. To amass creative energy to protect the Earth is as important to the next century as mass production has been to the last century [20th]. Its undeniable purpose will be the welfare of our planet and its inhabitants.

Although the company never truly materialized, the message did resonate. In that respect, we did succeed. I have heard over the years that my small business influenced some opinions, and I've been told more times than I need to hear that we were ahead of our time. Our message was actually behind the times. The decades leading up to the new millennium were inadequately responsive to climate change, and the risks were intensifying. We were losing valuable time, and that's what compelled me to dig deeper in my life; that's what spawned a new generation of activists. That's what drove Ken Ward and four others to shut down the flow of oil in the Northwest in 2016.

Today's protesters are ubiquitous and far greater in numbers. Governments have let them down as they have all of us when it comes to meeting the challenge of arresting climate change. If political leaders can't see the greatest threat to life in the twenty-first century, then those who were born into it will rightfully revolt against authority. This generation has learned also that there is an interconnectedness to racial, environmental, and economic injustice. That effect distinguishes modern protests from the Earth Day events of 1970. I have chosen to refrain from marching in protest during the COVID-19 pandemic that has swept America and the world. That hasn't stopped millions of others from protesting, but the dangers of community transmission of the virus keep me sidelined. I don't have the right to endanger someone else, and I'll be damned if I infect a family member who is at greater risk. Make no mistake, the protests of 2020 aren't going to die down easily until significant changes in our society are undertaken.

The virus transmission that is taking a heavier toll on Black, Latin, and Native Americans is a travesty. Having the privilege of visiting the Navajo Nation in 2012, I am particularly upset about the deadly spread of COVID-19 on their homeland. The gifts we keep *giving* Native Americans make them sick, figuratively and literally. So, no matter the motive for protesting, keep it up and hold those in power responsible for their actions.

Some of my fellow local climate activists contend that their sacrifices are moral choices. One, Kevin O'Reilly, a prolific writer and a distinguished history teacher for two dozen years, is making that case. He has given his talk entitled "Critical Thinking about Climate Change" for a year or more now.[10] The first note, the first slide, the first point, explains his motivation for climate activism. It is a photo image of his granddaughter. He cannot fathom doing nothing in the face of a climate crisis that will alter the experience of life for future generations, including hers.

I made a similar plea in the final pages of my first book on behalf of my two sons and two stepdaughters. I have no idea if they will make us grandparents but it doesn't change the fact that the planet our descendants inhabit after we are gone isn't going to be the same. I'm not confident enough to think that I deserve a greater share of natural resources than my children. If you do, please do not pass go, do not collect $200, and move directly back to the introductory chapter. The desire to give our offspring a better life than our own is ingrained in all of us. It is a sentiment as old as life itself. Not correcting climate change means that we no longer can control one aspect of providing for the next generation.

Even if we do not contain climate change to 2°C, which will be painful, we must all join in on the fight regardless. I also see the parallel with the stay-at-home orders of COVID-19 of 2020: we are sacrificing to save others. Whether you choose to have children or not, you still owe it to future generations to give them a fighting chance for a healthy existence.

I have bored holes in my son's heads since they were old enough to read the *Lorax* about my feelings on the environment.[11] I know they were already tired of my constant rants by the time they learned to talk back to me but I wanted to share my sentiments with them from early on. An early lesson was to equate the need to be kind with caring about the environment. It has held true in my life that kindness goes hand in hand with conservation. If you have a large environmental footprint, and can't make the correlation, it's time for some soul searching. Many of us make sacrifices knowing others choose not to. We do it because we care beyond our own generation in the hope that choices will still be available to them.

It is in that spirit that I find inspiration. Kevin O'Reilly understands that, and so does Fred Hopps, another local environmental activist here in my town. He maintains the oldest existing photovoltaic site in the country, the Dr. John H. Coleman Greenergy Park.[12] Fred, another former teacher, sacrifices his time to promote renewable energy and still gives tours of the solar field. He knows his efforts aren't always rewarded but he does it anyhow. He knows his contracting business suffers because of his volunteerism but he does it anyhow.

That is what motivates me: the willingness to sacrifice for the benefit of people we'll never know, and for the generations of children eager to live long healthy lives. And Fred doesn't even see it that way. He says, "It is my devoted cause so I can't even relate to the word, *sacrifice*." How does one person see conserving our habitat and fighting to mitigate climate change as a devoted cause while someone else pays no heed to the warnings? Within that question is the problem and the answer. Nothing short of a global movement will contain climate change. No one is immune from it, though the wealthiest of us may not feel the initial impact of environmental decline. If we activists can convince others that joining the cause is essential to life, then we might not have to glue ourselves to buildings or shut off valves. And if you still believe that lobbying corrupt government officials to enrich yourself and the fossil fuel industry is a worthy cause, it ain't just gonna be Kevin O'Reilly, Fred Hopps, and the rest of our merry band of local environmental activists protesting your efforts. Our sacrifice may include you, one way or another.

The word *reluctant* in the title of the documentary of Ken Ward and the other valve turners is poignant to me. When people face oppression, inequality, underrepresentation, denial, and despair, a reaction is inevitable. No one sets out to be a protester committing illegal acts of civil disobedience, but no one should stay on the sidelines when the status quo has been so inadequate in producing fundamental change. We all have been thrust reluctantly into the battle for the health of the planet. And we haven't won a lot of victories. With each loss is the growing anxiety to step up, march, lobby, and disrupt, if necessary.

The not-so-subtle message of the activists in *The Reluctant Radical* is that we need to end our damn reliance on fossil fuel energy as soon as possible. We can take other measures that will assist in slowing down climate change, but if we keep burning fossil fuels, we won't win the war. We have a long way to go here in the United States. As of 2019, we were still sourcing about 80 percent of our energy consumption from fossil fuels.[13] What drives someone to trespass, to march, to risk arrest? It is the recognition that the problem and the solution are on current paths that will never converge. Accepting the status quo means that we will never adequately fix this. That is what makes warriors out of mothers, knights out of children, and pipe turners out of the Ken Ward's. I didn't imagine I'd be out toting my sign, "Make America Cool Again—Stop Burning Fossil Fuels," either. I reluctantly used a ruler and it is legible—if I only knew in 1972.

Climate Justice and Protests of 2020

The state of the country in the summer of 2020 is markedly different from that of Earth Day 2018. Protests are ongoing and widespread, and divisiveness and dissolution are fueling them. It isn't lost on me that today's protests are not that dissimilar to the ones of the valve turners of 2016. Ken Ward knows that if we mitigate climate change, then the most financially hard hit of us will benefit. The least represented of us will be heard; the least responsible for climate chaos will feel safe. Climate justice includes ending discrimination. When the poorest of us are the ones in the crosshairs of climate change, then environmental justice means we need to look at the systems in place and make changes. It isn't the top wage earners who breathe the dirtiest air or are the first to flee floods and wildfires despite the fact that climate change doesn't discriminate. A system of oppression and lack of socioeconomic mobility has locked in this condition. The first waves of climate refugees aren't getting tax breaks or getting white-collar jobs.

Look at who was most directly impacted by Hurricane Katrina in 2005. These were poor communities—in poor states—that lacked the resources to prepare or in some cases even flee in the face of a climate disaster. It is estimated that 400,000 people became evacuees, and close to a thousand died as a result of Katrina.[1] The poorest of us deserve clean air, safe places of shelter, economic opportunity, and access to the same food supply. If environmental justice is to be attained then race plays a part in it.

The year 2020 represents a low point for this nation in addressing activism and protests. Arresting American citizens using their constitutional right to protest isn't the appropriate response of federal authorities or our police departments. Tear gas, rubber bullets, batons, flash grenades, unmarked vans, tossed bodies, and knees to the throat should be scenes from a war zone, not the streets of a developed nation. I have friends who proudly wear badges and despise these incidents as much as I do. They alone cannot fix the deeper problem of inequality and discrimination that have long festered within our society. Racial unrest was simmering when Trayvon Martin was murdered for his skin color. The injustice behind the law that protected his murderer represents a systemic failure. Watching Ahmaud Arbery being hunted down like prey eight years later means that we haven't improved the system at all. Throw in George Floyd's suffocation, and people of all races are out on the streets because they cannot tolerate one more scene of deadly discrimination playing out in America, land of the free.

And if you get a sense that I don't distinguish between peaceful protest, nonviolent acts of civil disobedience and looting and violent rioting, then you are incorrect. Nothing diminishes the potential for change, the good trouble, more than senseless violence and property destruction. All of the gain that can be achieved by peaceful demonstrations is eroded by irresponsible looting and rioting. Listen to me, you're doing your cause more harm than good by committing stupid, violent acts. It detracts from all of the good that can come from massive efforts like the Civil Rights Movement of the 1960s. I can't imagine what Martin Luther King Jr. would say fifty-two years after his death about today's racial unrest. He would be disappointed at the slow pace of change but more than willing to rise up again. We could sure use his strength and compassion at this particular moment in time. I'd say we're a long way from the mountaintop. And that was before January 6, 2021.

CHAPTER

Bonds Be Unbroken   2019

*I*n commemoration of the first Earth Day in 1970, I am paying tribute to this nation's settlers. No, not the European settlers, the Indigenous ones. Some of us are old enough to recall a 1971 Earth Day advertisement from Keep America Beautiful starring Iron Eyes Cody, who, in full Native American regalia, famously shed a tear over pollution. It remains one of the most influential advertisement campaigns of all time notwithstanding the fact that Iron Eyes Cody was an Italian American and the commercial was funded by companies opposing a bottle return policy.[1] All of America was originally settled, let's not forget that, including our current national parks. The decoupling of American Indians and their land was a long and protracted process of subjugation and deceitful negotiations. It's a stretch to use the term treaty to describe many of these transactions.

Western tribes like the Diné (Navajo), Lakota (Teton Sioux), Nimiipuu (Nez Perce), Nde (San Carlos Apache), and Tséhéstáno (Cheyenne), were forced from their territorial homelands after long, violent struggles. Conflict was inevitable, even likely, due to the vast amount of land covered by nomadic tribes. Now, 150 years later, their descendants are still fighting, but it's a different battle. They're fighting for respect, and some are losing.

Respect is a pillar of equality. The social, economic, and health disparity between the average American and Native Americans, including Inuits, is startling. Per the U.S. Census data for 2017, the median household income for Native Americans was about twenty thousand dollars less than that for the total population.[2] But that statistic tells only a part of the story. Compared with all other American ethnicities, Native Americans have the highest rates of unemployment, poverty, suicide, infant mortality, addiction, unsanitary and unsafe water, cardiovascular disease, smoking, diabetes, high school dropouts, alcoholism, and motor vehicular deaths.[3]

The Indian Gaming Regulatory Act (IGRA) of 1988 changed the course for some Indian Nations. The IGRA established regulations for sovereign nations to formalize gambling with the purpose of creating economic opportunities for tribal governments without negative external influence. Smaller tribes in territorial homelands within densely populated area, have prospered, including the Mohegan Nation, the Seminoles in Florida, the Mashantucket Pequots in Connecticut, and the Mdewakanton Sioux in Minnesota. In some cases, these casinos have breathed new life into depressed reservation economies.

Some of us have developed, designed, or furnished Native American casinos throughout the country, and even more of us visited, dined, or gambled at them based on recent gaming statistics. According to the National Indian Gaming Commission, revenues reached thirty-two billion dollars in 2017.[4] These figures eclipse gambling numbers from the Vegas Strip by a wide margin. Although they are regulated differently, there is no question that gambling is big business. Ever since 1988, Native Americans have had a hand in the game, so to speak.

I often wonder what their brave ancestors would think of their tribal lands becoming enormous halls filled with neon-lit slot machines and non-Natives in costume. One Lakota I spoke to, Alex White Plume, is opposed to the idea of American Indian casinos. He sees it as another effect of colonialization of Native Americans and he says that the monetary benefits don't equate to infrastructure needs like road and housing repairs that are much needed on many reservations. He agreed with me that what Native Americans need is more respect and greater appreciation for their way of life and for people to visit and support their cultural activities and businesses. Perhaps, most importantly, Alex suggested that non-Natives should recognize that forced assimilation brought trauma to many proud Indian Nations and they are struggling with that concept still. American Indian casinos and gaming centers are not a logical sequence in the evolution of Native Americans. I suspect the concept is a divisive one for most tribes. And although Alex White Plume is not a fan of Indian gaming, it remains a source of money for a number of tribes. In the end it may not represent the story of their people but hell it may be the closest thing to remunerations they'll ever get from us.

As we approach fifty years of Earth Day celebrations, let's do what we can to show some respect for our original settlers. Trading posts were the first buy-local markets, and there are many Native American-owned businesses on reservations to this day. Economic equality shouldn't be so dependent on gaming.

And what about those reservations established for larger Indian Nations? With or without casinos, they need our help. The Navajo People have a 42 percent unemployment rate. Close to half of Alex's Lakota People on Pine Ridge live below the poverty line. The leading cause of death for San Carlos Apache children between fifteen and nineteen years old is suicide, and for their parents it's alcoholic liver disease.[5] A Northern Cheyenne on average lives twenty fewer years than the rest of us.

Our hospitality industry is a well-connected, supportive community of members. I am fortunate to work in such a tight-knit industry. I've been promoting sustainable practices within this industry for close to twenty years, and for the past ten I have been writing Earth Day articles. Since then, the stakes have gotten a lot higher. The Intergovernmental Panel on Climate Change has advised us to limit further global warming to 1.5°C to avoid climate chaos.

Sustainability is a mission that many of us have adopted, and it's no longer a side function of hospitality practices either. But sustainability is like another broken treaty if the poorest of us are the hardest hit. The recent bomb cyclone devastated Alex's folks on Pine Ridge. Thousands were stranded without drinking water. Lives were lost. Other reservations in the path of the storm had similar experiences—places already in the clutches of economic despair.

There is space for our original settlers outside of the physical and cultural barriers that constrain us. Visit their traditional homelands, attend their cultural events, purchase their handmade goods where possible, and drop a few dollars at the casino. Earth Day is one day in the spirit of goodwill but sustainability is a commitment to all days and all people, and a homeland where they gracefully intersect.

Vanquish, Borders and Reservations

Physical and artificial borders represent failures of imagination and design. They don't encourage the sharing of ideas. The exchange of knowledge and culture is a sign of stability and progress. When that free flow of goodwill and acceptance ceases, negative consequences ensue. We often behave poorly when we are driven by fear, and we become susceptible to destructive instincts. There are many instances where cultures collide because of a fundamental break in mutual understanding and trust.

If we needed any other signs that fences, borders, reservations haven't worked for a specific race of people, we don't have to look far. And, by the way, who has the right to ask another human being to assimilate to a culture that isn't their own? This is our sad history here in the United States toward Native Americans since our first landings on this side of the Atlantic.

The nineteenth century, acclaimed as the settling of the West, also proved to be a period of ambition, cultural destruction, distrust, and deceit. The reckless slaughter of millions of American bison, which was a government policy to starve out any remaining free-roaming Plains Indians, was not only an act of ecological stupidity but an act of pure cruelty toward both. Our Plains Indian ancestors were some of the last holdouts, proudly standing up to the onslaught of a race that invaded them like a swarm of locusts.

I began reading about the history of American Indians when I was quite young. From *Bury My Heart at Wounded Knee* to today's *Flowers of the Killer Moon*, my life has been consumed by this historical wrongdoing.[6,7] We have a long way to go in achieving equality among White and Black Americans but just imagine in the nineteenth century if we took the initiative to banish freed Black Americans to a colonized Liberia as was proposed? It would be perfectly clear what manipulated destiny means, and the pain of capitulation, captivity, and handouts.

Native Americans do share the horrors of enslavement with African Americans, if not in numbers. To think that we still honor Christopher Columbus with a holiday in view of his crimes against the Taino and Caribs, including abduction and wanton slaughter, is beyond insensitive.[8] In fact, the number of Natives once enslaved will truly never be known.

Three hundred fifty years after Columbus's misguided journey, Native Americans in the United States found themselves surrounded, unwanted, misunderstood with odds against them also. Many tribes were decimated by disease just like the Indian Nations of the Caribbean were by their visitors. Others were officially deceived into signing away land, and yet others succumbed without bloodshed.

Henry Dawes, U.S. congressman and senator from Massachusetts of the late 1800s, authored legislation to parcel allotments of reservati0n lands to individuals and households of the various Nations. The Dawes General Allotment Act of 1887 was enacted, ostensibly to provide American Indians with acreage to farm and raise cattle and achieve the prosperity of White Americans. This ill-conceived law did the reverse. Within forty years, Natives lost more than half of their designated homelands.[9]

Notwithstanding the fact that many Plains Indian Nations were nomadic, sedentary agrarian lifestyles were foreign to their culture and religion. Penned in, subdued, and shadowed by a race of conquerors that made little attempt to understand a different way of life resulted in a complete policy failure. Included in this policy was the onslaught of representatives from all White man religions to convert the "savages" from practicing their own unique religions. The ban of the Sun Dance, which was an important ritual in the development of American Indian youth, was particularly cruel. This hollowing out of the soul of many Native American tribes is directly correlative to the raw statistics referenced previously.

The final indignity for many came in the form of American Indian schools like the Carlisle Indian School. These preparatory schools were, in theory, institutions created to teach Native American children the skills they would need to be productive in Euro-American society. Don't kid yourselves into thinking this was a preppy man's boarding school for Native Americans. These children were forcibly plucked from reservations across the country, removed from the only culture and family they'd ever known, prohibited from wearing traditional clothing and hairstyles, prevented from speaking their native languages, and in some cases, placed in locations with temperatures and conditions so distinct from their own as to be deadly.[10] Even if they survived these institutions, their return to tribal reservations meant that their cultural identities would be stripped of them. Not white, not Native, but somewhere in between in a yet-to-be-determined identity known only to them.[11] Learning English shouldn't have had to come at such a steep price. This poem underscores much of this:

Manipulated Destiny

We bestowed upon them the Dawes Act,

and dared to use the phrase "Indian giver."

Sequestering Natives—-a form of treatment known
to gray wolves and grizzly bears.

Plains warriors resigned to plows, hoes and spades,
counting seed not coup,

watching the exodus of the buffalo by the lights of
the Union Pacific.

We sheared the bangs of their children in Carlisle,
Pennsylvania

so they could look like their conquerors.

We banned their Sun Dance to save them from hell.

Alcoholism, greed, disease and depression,

we've taught them enough, I think.

But I want to ride breechcloth on bareback,
shirtless and sun-drenched

ranging boundless across the unmilked bosom
of the Prairie.

I want to dance beneath virgin skies uncorrupted
by city lights,

dwell in nomadic villages that leave no footprint

on land that has no price.

Be quenched by rivers yet to run red.

Who will teach me?

(2006)

With such insurmountable odds, heroic last stands became the sad ending to the last free Native Americans: from Chief Tecumseh at the Battle of the Thames to Chief Joseph's Nez Perce in the snow just miles away from freedom at the Canadian border, from Crazy Horse's assassination at Pine Ridge to Victorio's doomed defense at Tres Castillos, from the frozen mangled corpse of Spotted Elk at Wounded Knee to the suicidal prison plunge of Satanta in Southern Texas, there are far too many Native American tragedies that have shaped our recent history. It is still fresh in the minds of many Native Americans. It certainly was when I interviewed Peter Jemison of the Seneca Nation in *One Green Deed Spawns Another*, and it sure was when I spoke to Alex White Plume.[12] They still expect European Americans (non-Native Americans) to live up to their treaties and promises. And they will continue to lobby to see that we do. Many tribes have suffered harshly due to borders, conquest, and relocation. And, think, we couldn't even correct

Christopher Columbus's geographically confused term *Indians*, for Native Americans, over five hundred years later?

Alex White Plume told me that Native Americans suffered from assimilation shock and historical trauma as a result of the institutionalization of their race. And despite the cultural revival of many Indian Nations, they are still seeking equal respect for their languages and ceremonies. Some, of course, are using the courts to pursue the return of lands taken as a result of broken treaties, including the Lakota. In a case of inexplicable absurdity, Alex told me that they have to get a permit to go into the Black Hills despite the fact that the Supreme Court in 1980 ruled that it was illegally obtained from the Lakota and offered them a $106-million-dollar settlement, which they refused.[13] Alex also opined that traditional Native languages and ceremonies are as real as all other cultures and deserve to be honored for perpetuity.[14]

The 1970s were an era of violence for the Lakota. The reign of terror following 1973's occupation of Wounded Knee by traditionalists and members of the American Indian Movement (AIM) was waged by the tribal government of Richard Wilson against traditional Lakota on Pine Ridge. What followed was the lawless persecution of the Lakota by Wilson's makeshift army called the GOON squad. This reign of terror included countless unsolved murders and the incessant harassment of traditionalists. Alex's late wife, Debra, was a recipient of this harassment and was wounded from bullets targeting her family. The tribal police acted in conjunction with the U.S. government still spoiling for a fight from AIM's '72 occupation of the D.C. offices of the Bureau of Indian Affairs and of Wounded Knee months later. This chapter in history is spelled out in Peter Matthiessen's gut-wrenching book, *In the Spirit of Crazy Horse*.[15]

If you don't believe me that the U.S. government hasn't forgiven the Lakota, just consider that they have incarcerated Leonard Peltier, a member of AIM, forty-five years and counting, for the death of two FBI agents, Jack Coler and Ron Williams, in 1975. The onus for the agents' deaths belongs at the feet of Richard "Dick" Wilson and his law "by any means necessary" policies. That he was given the active support of federal agents demonstrates their culpability as well.[16] Meanwhile Leonard Peltier rots in prison despite the fact that his trial was a farce and that the evidence to convict him was beyond dubious, including coerced and false testimony. Of note, numerous other prisoners who were tried justly have recently had their sentences commuted, which means that the Peltier case is now an act of inhumanity.

Respect

Alex White Plume believes that his Indian Nation deserves respect and that they have suffered enough historical trauma and grief. Furthermore, he believes that the children of his people deserve liberation and a general cleansing as a result.[17] For that reason, Alex was one of the organizers of the Big Foot Ride in 1986, which traces the path that Si Tanka, interpreted as Spotted Elk and indelicately called Big Foot by White men in the nineteenth century, took to escape the pursuing U.S. Cavalry in 1890. Spotted Elk, after accepting and accommodating the Hunkpapa Lakota Indians that had fled the Standing Rock reservation after the arrest and murder of Sitting Bull, determined to take his band south to Pine Ridge. Spotted Elk, who was known as a negotiator and peacemaker, was honoring the request of the Oglala Lakota to assist them in resolving an intra-tribal matter. Spotted Elk, his fellow Miniconjou Lakota, and the Hunkpapa Lakota followers of Sitting Bull made that fateful journey that culminated in the massacre at Wounded Knee on December 29th. This incident will forever be remembered as one of the most tragic in the annals of conflict between Euro-Americans and Native Americans. The wanton slaughter of mostly unarmed men, women, and children by the U.S. Cavalry at Wounded Knee is a blight on American military history that can never be removed.

I have been to the mass grave at Wounded Knee. It was emotional for me having known the story for years. It is not much more than an extended slab of concrete and a large upright headstone encased in a chain link fence. A few of the victims are listed on the headstone, representing but a small portion of the total buried there. No one knows for sure the number of victims of the massacre other than it exceeds the losses of U.S. soldiers at the Battle of the Little Bighorn. The Little Bighorn Battlefield National Monument is a part of the National Park Service; Wounded Knee is very clearly not. To me, it is eerie and unfinished, and it will always be a marker for our misguided policies toward Native Americans for centuries.

Alex and his brother along with seventeen other Lakota determined to retrace the steps that Spotted Elk took to Wounded Knee as an act of honor and healing. The Big Foot Ride is now in its thirty-fourth year. These proud descendants make the long trip, near two hundred miles, on horseback to commemorate the original trip of 1890. In many ways the journey is as much a cleansing experience as it is historic, according to Alex. Perhaps it is also a means of purifying a people who have known an imbalanced share of loss.

Alex White Plume, former president of the Oglala Lakota Nation, is also well known for his commitment to industrial hemp farming and his interactions with the U.S. government. As a traditional farmer on the Pine Ridge reservation, he has tried his hand at a range of crops. But it was his interest in growing low-THC (tetrahydrocannabinol) hemp for fiber that attracted the attention of the Feds. Without delving too deeply into the trials of returning hemp cultivation to the United States and its messy, overregulated past, suffice it to say, few crops have ever received such scrutiny. My good friend and Alex's, Barbara Filippone, a hemp fiber design and processing developer, has bristled at the lazy comparisons of hemp and marijuana. She has described the confusion as not understanding the difference between a bell pepper and a chili pepper—and she is convinced that the Drug Enforcement

Agency (DEA), the agency responsible for its management, has long known the difference but stubbornly refuses to update its policies. Today, as the distinction between cannabis and low-THC hemp gets blurred by the media and health "gurus," a productive crop like low-THC, long bast fiber remains mostly idle. Alex might be producing fiber for rope, apparel, plant-based food additives, hempseed oil supplements, and other industrial hemp products, if not for the persecution by the DEA.

Pine Ridge, like all Native American reservations, is recognized as sovereign land. Alex and his wife, Debra, tested this out and were reminded that White men don't always abide by treaty rights. They planted hemp on several acres of family land from 2000 to 2003. All his hemp crops were cut and seized before harvest by DEA officials, which placed Alex and his extended family in dire financial shape.[18] For a place and people of restricted means to lose economic opportunity is both reprehensible and racist. Many Indian Nations are still seeking restitution from broken treaties. In Alex and Debra White Plume's case, honoring sovereign rights guaranteed by treaties was tested once again, and unsurprisingly, the U.S. government failed to abide by its obligation to a signed treaty—specifically, the infamous Ft. Laramie Treaty of 1868, which ceded the Black Hills to the Lakota, well, that is until General Custer's expedition found gold there six years later. Within this treaty also is the guarantee of agricultural development rights that Alex took at its word.[19] In fact, much of the negotiation on behalf of the government was to strongly encourage farming practices versus nomadic activities like buffalo hunting for example. Alex lost a lot fighting U.S. authorities a hundred and fifty years after the Ft. Laramie Treaty.

It took close to twenty years after Alex's battle with the DEA for the legal growth of hemp to return to the United States and to the Pine Ridge Reservation, but that hardly absolves our government of its harsh treatment of the White Plumes. It serves as yet another example of discrimination and prejudice. Alex has every

right to be bitter but when we spoke by phone, he was anything
but so. He has since lost his partner, Debra, who passed away in
late 2020.[20] She was every much the warrior of proud Lakota
tradition, having dedicated her life to protecting her people's rights
and culture.

Restitution

Alex and many proud traditional Oglala Lakota continue to reside
on Pine Ridge, which is also subjected to the conflicts between
American Indian casinos and cultural integrity. Some Indian
Nations have found the gaming industry to their benefit, others
not so. The larger of these Indian Nations and Reservations, in
many cases, are not close enough to urban centers to host mam-
moth casinos. That shouldn't have any correlation to poverty, but
sadly it mostly does. Economic despair has stalked the Lakota
and other tribes. Economic inequity leads to countless negative
outcomes. There are too many miles among these proud people
to ride down that despair.

We can start by seeking business opportunities with Native
Americans represented in the United States and support Native-
owned businesses. The Department of the Interior, led by a Native
American for the first time, has programs and funds to support
economic development on Indian Reservations. Consider how your
business can include Native goods or services in your products or
supply chain. Does your company have the ability to apprentice and
train skills on a reservation? Most reservations qualify as HUBZone
areas and have preferential access to federal contracts and services
as well.[21] Can your business or industry bring back some aspect of
outsourced manufacturing with the help of a federal program to
create jobs on a reservation? And, if you are not in a business or
service that could benefit from Native American labor or compa-
nies, visiting and spending money on a reservation can help. If we

can raise millions of dollars to build a wall to keep Mexicans and Central Americans out, then we can raise money for the original Americans who welcomed us in. And shop Native as well as local.

The subjugation of our Native brothers for hundreds of years parallels what we have done to nature in far less time. Our future is in the redemption of their future, one that we can recreate together after five centuries of missteps. We are at risk of losing the last interpreters of Indigenous customs and languages. And by doing so, we may never understand their deeper spiritual connection to nature, one that saw our world as a benevolent deity versus the tapped-out and exploited wellspring of hydrocarbons that we worship today.

Maybe we will turn the page and seek a more reverent relationship with our natural world before climatic changes overwhelm us. Maybe some retrospection is necessary in the case of Native Americans, too. Perhaps it's time to teach Native American culture and history in our classrooms versus celebrating with a day off every second Monday of October. It may help us appreciate

the importance of diversity and integration and the gifts of a living, benevolent planet. Hell, who knows, maybe it will just close a bunch of ugly chapters in a story. The more important question for today should be, will we choose to help liberate, bolster, and promote Native Americans, or will we continue to suborn and resist their ascension? It remains to be seen. And, in the spirit of Leonard Peltier, let the man go, damn it.

12

CHAPTER

Earth Day   2020

*T*he knack of any writer is to not over sentimentalize the passing of time and lives. But at certain points, we all do. I embarked on this book project years ago with the intent of completing it by the fiftieth anniversary of Earth Day. After ten years of Earth Day essays, I set April 22, 2020, as a target date of publication. I knew it would be a different venture than my previous book, which had no set timeline. As a chaotic writer, deadlines can be a challenge, and my love for short stories and poetry has always worked against me as a nonfiction writer. Coupled with a global disease that has halted the world in its tracks, I failed to meet my deadline. Just a note, if it turns out to be a live horseshoe bat that originated COVID-19, it should put a permanent end to any open wet markets where species are forced into dense *interaction. Believe me, a stressed-out bat is just the opposite of a crime-fighting caped crusader.*

Reflection

In the spring of 2019, my father started to weaken. As I noted in the acknowledgments in *One Green Deed Spawns Another*, Dad and Mom were my first and fallback editors. Dad authored near a dozen books including bringing to light Union Generals and Union Regiments from the Civil War that had disappeared into oblivion. He also authored books on teaching social studies for high schools, and he was the primary author of the history of SUNY Geneseo (New York) published in 2008, and lastly, wrote a biography of his mother's family.[1] He was an honored college professor during his career but moreover he was an attentive and loving father to two sons. Despite that love, I veered off into an abyss during periods of my life. Like many of us, I learned to accept certain things about myself, and ultimately, the path I was on, which never followed a straight line. Both Mom and Dad projected big figures as models for me as a young man. It took many years for me to comprehend that the work they both put in to reach these goals had nothing to do with dreams or utopian ideals. Both of them were achievers and unified in their commitment to a better life for their children. My shadow isn't that big in comparison.

I knew matters had worsened in Dad's life by July, but on August 11, 2019, we were told by his oncologist, Dr. Casulo, that she had successfully treated Dad's mantle cell lymphoma but that a different terminal cancer was growing aggressively in his upper lung. Although she could not treat the latest of his four cancers, she kindly put him in contact with a colleague who specialized in this particular cancer. My father, in his inimitable style, thanked her for taking care of him for five years, and upon learning that he had yet another deadly cancer diagnosis said, "Well that's a kick in the pants."

Reaction

The kick I felt still churns today. I know others can compartmentalize tragic events, but I found that many days and nights of 2019 were consumed with thoughts of losing Dad. Being an independent contractor, I could take time as needed to be with him but writing this book became a distraction. He was an accomplished lecturer and speaker, and to his credit was asked to speak at a number of funeral services and celebrations of life. In fact, my grandmother requested before she died that he do so for her—and she was his mother-in-law! He made us all realize the importance of communication—especially poignant to me as a husband, father, and stepfather.

Well known to close friends of ours is that he mailed a family letter every Sunday for close to forty years. Without chronologists we lose track of more than time. Life's events slip easily through our memories, people fade from view, and recall becomes skewed without the written word. Dad made sure I have the outline of most of my life in four massive black binders.

I know that my father had faults and I'm not blind to them but he was big on making amends. I haven't truly gained that gift. Regardless, it was a privilege to honor him in front of an overflowing and esteemed gathering at his celebration of life service. I was additionally moved by the scholarship award that SUNY Geneseo established in his name, formally acknowledging his contributions to secondary education at the college made his family forever proud.

Both Mom and Dad taught me the basic concepts of ecology and sustainability in a number of ways. Neither of them would compare their actions to the pioneers of the modern-day movement but they both took conservation seriously. Mom read books to her library classes recognizing Earth Day and nature. As they both did with most lessons outside of the classroom, they showed us by example, including valuable lessons of wasteful consumption. For one example, Dad found someone to repair almost everything he owned, from watches, typewriters, grandfather clocks, office chairs, canvas tents to lawnmowers, air conditioners, and a most identifiable item: penny loafers. He had shoe heels repaired so many times that he far outspent in repairs the original shoe price. I later learned the hard way about his feet. Genetic traits aren't always a gift. His supinated step wore down most of his shoes prematurely, and mine may be even more severe. My gait is not only supinated but my feet are a B width. I'm lucky I can walk at all.

My father died on November 7, 2019, five days before my fifty-seventh birthday, and six before my eldest son's twenty-fifth. I'm still trying to understand what it means to be present at a loved one's last breath. I was holding one of Dad's hands and Mom had the other one when he died. He stopped breathing and turned gray so fast as if something out of a Stephen King book, a screenshot of the burned-out chassis of a six-foot-two-inch, physically active man. I was still holding his hand when the hospice nurse called the funeral home. And that is the physical end of my father. I am not supposed to tell you that my heart was broken but I can't exclude it either because with loss comes resolve, and in his memory, I found the inspiration to continue to write.

Near his last weeks he lamented the planet he would soon leave to his children and grandchildren. He had outlived his body; he didn't outlive his senses or his worth. He promoted the science of climate change as you would expect from an educator. He knew that his lamentations wouldn't startle me or seem hyperbolic to someone keenly aware of the issues. He was intrigued by the many technological achievements in renewable energy and conservation. We often sent back and forth articles, Internet site links, quotes, and profiles of environment leaders, and debated climate related problems for twenty years or more.

Surprisingly, he was not an optimist despite his very positive public demeanor. He was much more of a realist. Dad cheered on my efforts, including the protest signs, marches, and activism, and often shared pics of us at events somewhere. And, as much as I had researched and studied to earn an MBA in Sustainability, it wasn't unique to me. He was well read when it came to the environment. It pleases me that he did get to see most of the Earth Day essays in this book.

One final reminder of Dad's (and Mom's) effect was the inevitable task of assessing the worth of one's lifetime of "stuff." Some of it is just stuff but Dad was always thoughtful and at one point we discovered a small index-card notebook. Reviewing it, we realized it was a handwritten check register of all his charitable donations. Dad and Mom were educators before their retirements; they lived middle class lives and never lavished themselves with excesses. Their money was very well managed and Mom is in good financial shape as she moves on without him. Additionally, my mother continues the tradition of charitable donations today. Looking closer at the notebook, it dawned on me that they had donated thousands of dollars to progressive organizations, many of them doing the noble work of combating climate change, preserving wildlife, conserving biodiverse regions, saving sea turtles, planting trees, and supporting climate leaders in politics for decades. Sharing their modest earnings seems so trivial until you see the amount and consistency. I never got to ask Dad why he did it. I know many people with far more means, and with equal environmental concern, who choose not to donate. It may have to do with his always championing the underdog or maybe it was a lesson for the rest of us. I mean, damn, he could have bought new shoes.

We gain valuable lessons from those who lead by conscience. We could use some at this moment in time. We were as polarized in April 1970 as we are at the close of 2020, and we are suffering through a number of crises simultaneously. Somehow, years ago, we mostly put aside our politics of the day and concluded that Mother Earth deserved honor and protection. We must do so again with even greater zeal. I think about what Senator Gaylord Nelson would demand of us in the face of today's climate chaos. I'm pretty certain he wouldn't be satisfied with a few teach-ins. He would be leading us by action. We owe it to Senator Nelson and Frances Moore Lappé; we owe it to Shawn Cantrell of Defenders of Wildlife. We owe it to all of today's youthful activists, and to others whose voices have been silenced permanently like Homero Gómez González and Raúl Hernández Romero. We owe it to Bob and Doris McGrath, and to the Clarks on Homeland Farm. We owe it to Marshall Saunders, and to Jim Mulloy and Rob Bonney, and we owe it to Alex and Debra White Plume and to Ken Ward. We owe it to science, and the scientists, George M. Woodwell, Max Holmes, Peter C.H. Pritchard, Mark Westneat, and Mark Dorfman. Moreover, we owe it to the wolves of Yellowstone, that curious orangutan in Borneo, to Myrtle the sea turtle, and all of the diverse fauna and flora that co-occupy this planet. And I, to Wayne Mahood.

Amends

I finished an award speech in 2011 with this environmental plea for help and on this fifty-first anniversary of Earth Day, I'm going to ask again.

To make freshwater free, clean air communal.

To cool our atmospheric coils.

To repaint our reefs, restock our streams.

To keep the stripes on tigers, the paddles on sea cows.

To keep ice on the ground, birds in the sky.

To offset not upset.

To reduce, reuse, recycle,

but mostly react.

I might add:

To see that ancient, native, and tropical forests stand free.

To decarbonize our dwellings and destinations.

To shift our diets from factories to small farms.

To ramp up renewable energy and ecological innovation.

To stop human detritus from sickening our oceans.

To sacrifice to love more,

and pay it forever forward.

Earth is a rotating orb of wonder, the only home we'll ever know. It was a random concoction of the elements that sparked this existence of everything. A living planet is the omnipotent gift we were granted, but we have since learned that it may also be fragile and subject to exploitation. How we elect to mitigate any pattern of decline will be our greatest achievement or our most profound failure as the dominant species of the Cenozoic era. The chances of our many endangered species retaining a foothold on Planet Earth rests in the balance. And the loss of biodiversity is in itself a sign of declination. How do we respect life in simpler, more fundamental ways that are commensurate with the urgency it demands? In my opinion, the intricacies of sharing it will prove to be our species' defining acts. Will we commune as gentle giants or will we be remembered as conquerors? This question remains unanswered but it isn't one we should ponder too long. I believe we'll make amends; we often do as a trait of our species.

Embrace this peculiar experience for it's all we'll ever know both coming and going. And preserve it, not in fear, but in hope and conviction. If we discovered at an early age that life exists beyond humans then we lack excuses for inaction now. Let's make *E* stand for Earth. Let's make *I* stand for infinite. Let's make *U* stand for universal, and let's restore the faith of that little fourth grader by righting the *Last Wrong Dictionary*.

EPILOGUE

⌁

What Does Earth Day Mean to Today's Youth?

My sincere thanks to Rebecca Marasco, Gretchen Marks Crane, and others, for compiling these responses.

[Author note: lightly edited to maintain original voice]

Mallory, 23, New York

To me, Earth Day is not only a day to celebrate and appreciate the Earth, but it is a day of awareness. With the emerging global climate crisis we are in, now more than ever is it important to educate others and make them aware of (a) what is happening to our planet and will continue to happen if we keep living the way we are and (b) what we can do to change that projection. Earth Day is a day for us to come together as a human race and advocate something we all love—our home, planet Earth.

Zachary, 25, Massachusetts

Earth Day is a way to really think about the Earth and what it means to you and appreciate the natural world to get down to what's important.

Robbie, 9, New York

What Earth Day means to me is great. Because people pick up litter and I think we can.

Andy, 30, Massachusetts

Earth Day is a reminder not to take our planet and its many resources for granted. A day to inspire us to give back to this little blue dot we all call home.

John, 36, New York

My sophomore year in high school I was picked by my wildlife bio teacher to go to a local grade school to make Earth Day stuff . . . pinecones with peanut butter and bird seed, planting seeds in solo cups to take home. It made me light up and I saw the kids light up too. Was my first interaction as a teacher or facilitator of an activity that I felt meant something.

Tessa, 9, New York

For me Earth Day means that you have to help out the Earth. That means pick up all of the garbage, like picking up all of the garbage on the land and in the water, such as lakes, ponds, and oceans. So, if you have those plastic things from soda bottles make sure you cut them up or else they will end on a sea turtle's neck. That is what Earth Day means to me.

Dan, 32, North Carolina

The Lorax.

Christina, 35, Maryland

To me, Earth Day meant getting a tiny sapling tree at school, taking it home and planting it. Then, having the tree not survive because it was too tiny.

Nicolas, 30, Connecticut

We all celebrate holidays to positively reinforce what we value, right? Earth Day is this giant blue green ball. This home that unites us. Many still take the chance or excuse to plant a tree, pick up some garbage, and do whatever works for them to feel good about the planet. Sure, we have a lot of work cut out for us to fix it all. *Every little drop in the bucket helps fill it up. Thanks, Earth Day, for getting so many drops in the bucket.*

Maddie, 29, Rhode Island

A reminder about where we came from and how special the Earth is. A reminder to take care of where you came from. Not just on Earth Day but every day.

Christopher, 22, North Carolina

Earth Day, to me, means taking the time to appreciate our planet and understand the effect we can have on it, both negatively and positively. There's only one Earth, so let's take care of it.

Mary, 10, New York

Earth Day means to take care of the Earth and be nice to the planet. Like watering plants, planting trees and different types of plants. Just doing something to help planet Earth.

Jordan, 26, Connecticut

It means we decided to make a holiday instead of actually doing some-thing about climate change.

Jack, 23, Massachusetts

Earth Day to me means a day where we take a step back and think about the Earth and what it means on the individual level and the society level. How our relationship to the Earth is fundamental and how we see ourselves as humans gets us thinking about how we can improve our relationship in the future.

Christopher, 39, Pennsylvania

Earth Day means to me clean sailing. Letting the wind and the sea that we should be protecting, take us where we want to go.

Cecilia, 10, New York

Earth Day means to me that people help the world and clean it up and not have junk and plastic all around the world because if any animals eat that it can kill them and animals in the water can get stuck in it and it can cut their circulation off and that would not be good so Earth Day is cleaning up the world to me and making it a better place to live.

Bruno, 25, Massachusetts

I can easily say that Earth Day means more to me than Christmas. In my opinion it actually should be for everyone. On Earth Day I try to reflect how Earth is a living being and how I can be a part of its healing system.

Damon, 29, Colorado

Earth Day to me is a time for everyone to be present with Mother Nature. To acknowledge what impact humans have on Her.

Brody, 10, New York

What Earth Day means to me is that nobody should litter because if one six-pack ring goes into the ocean, then some penguins are going to get stuck on a ring and will probably die.

Brett, 17, Georgia

Earth Day is a time where everyone can forget their problems and enjoy the beautiful Mother Nature that surrounds us.

Katie, 30, New York

Earth Day is a time to celebrate Mother Earth and all that she provides. It's a time to reflect on how we can better co create with the Earth instead of destroying it.

Carson, 10, New York

What Earth Day means to me is it is a day where we clean up by picking up garbage, cleaning your house, etc. Then when you are done cleaning up you can go to a park and help pick up trash there. Then you can help anyone clean up that you know. That is what Earth Day means to me.

Francesca, 9, New York

I think Earth Day means to send peace to the Earth and to make sure you mourn the Earth. I think you should mourn the Earth because [it] does a lot for you and everyone else. Also you should always be respectful to the Earth no matter what.

Anita, 38, Connecticut

It is a day to reflect on our actions and their effects on other species and the climate. It is a day to remind others to reflect as well and bring behaviors to light that should be changed.

Noah, 21, New York

To me, Earth Day means taking action and accountability to make our planet more sustainable and healthier. It means coming together as a global community and putting in the work that is necessary in order to better this Earth that we all share.

Rachel, 9, New York

Earth Day means when people help out with picking up the trash and helping to spread kindness around the world helping people do better. Redoing the world like refreshing it, doing caring and doing kindness. We are all helping each other spread caring in the world—the whole world.

Samantha, 26, Washington

To me, Earth Day is both motivating and depressing. It's a reminder of the terrible state of the world's environment and the crises that will increase and that many are already experiencing due to climate change. But it's also a rallying cry. There are people who care deeply about turning things around, myself included. Now we just need the courage and will to keep pushing. Earth Day and every day, I try to remind myself of that.

Kelsey, 31, Rhode Island

Earth Day to me means a time when we can reflect on what we can do for our planet in regards to preserving it and enjoying all the gifts it has to offer us.

Armando, 9, New York

What Earth Day means to me is that it is a day to celebrate what Earth has to offer to everyone. Another thing is that it is a day to help Earth and to go outside and play and have fun in nature.

Ryland, 10, New York

Earth Day means to me that a single day could be motivating enough to save us [and] to think about global warming and save Earth because as my mom said the longer you leave it the worse it will get. So I think it is a special day of the year that is good.

Gregory, 29, Rhode Island

Earth Day is like a reminder time for me to think of our round environment here. To remember that we are on something. It's a finite piece of mass. Earth Day for me is a day to remember that and connect with it. Maybe touch the ground or plant a tree.

Magali, 31, Lyon, France

Earth Day should be every day. It's the respect for the planet we live on. All these pieces we are sharing with, we should all be in harmony

Charlie, 10, New York

Earth Day means to me an entire day to celebrate the planet we walk on and not take for granted. It, to me, means focusing on the Earth and all its quirks and embracing said quirks. Also, Earth Day, again to me, means to love and keep our planet healthy because to everyone on Earth's surface, it's like our pet and we want to keep it healthy 'cause it's the only Earth we've got. That's what Earth Day means to me.

Liv, 30, New York

Earth Day is like a New Year's resolution. Could be a great catalyst for change, but likely just a few hours of good intention.

James, 25, Massachusetts

Earth Day to me means a day to reflect upon my impact on the Earth and use of, I suppose. All the things that the Earth gives me but that I don't really reciprocate for, I suppose. It's a time to reflect and also think of how we can do better as opposed to endlessly taking.

Tayla, 9, New York

I think Earth Day is about taking care of the Earth, for example picking up garbage and watering plants. Taking care of wild animals, feeding them and cleaning up their homes just to help out. And most importantly is the water helping fish with taking out fishing lines and also [protecting] turtles.

Andrew, 29, Rhode Island

Earth Day means to me a day that is supposed to be a celebration and recognition of we as human and all that that makes up Earth and everything that is combined in the environment, and the atmosphere we live in. In recognition of this is what we have, we have to protect what is our home. To celebrate the joy we find in life and the joy we find on this planet.

Kevin, 33, New York

Earth Day is a stark reminder of the quote "We do not inherit the earth from our ancestors, we borrow it from our children." We must act accordingly.

Jessica, 31, Rhode Island

Earth Day means a day when we appreciate the Earth by picking up trash and meditating with the Earth and being grateful for her and all of her creations.

Micah, 9, New York

Earth Day to me means that we celebrate the creation of our world, Earth. So when Earth Day comes I try to dress up as grass to celebrate the creation of the environment and the world.

Pearce, 29, Colorado

Earth Day feels like something you do because it's trendy. Climate change is too important to just be a day on the calendar.

Ashley, 29, Rhode Island

Earth Day means a highlighted day in which we are supposed to draw our awareness and actions toward benefitting Mother Nature, Gaia, our planet with which we are all intrinsically connected . . . but Earth Day should be a reminder of every day—serve as an alert to bring conscious awareness to those who are blissfully ignorant to the world and connection we have with it, or have simply forgotten that WE are all a part of the life of this planet . . . we must respect and care for our world by starting at the microcosm of caring for each other and our environment through thought to ideas to perspectives to actions to the very reality we create!

CHAPTER NOTES

꙳

Chapter 1

1. Elizabeth Kolbert, *The Sixth Extinction: An Unnatural History* (New York: Picador, 2014), 261.

2. Daniel T. Cross, "IUCN: Nearly 30,000 Species Are Facing the Risk of Extinction," *Sustainability Times*, July 22, 2019, accessed August 9, 2019, https://www.sustainability-times.com/environmental-protection/iucn-nearly-30000-species-are-facing-the-risk-of-extinction/.

3. MercoPress, "Thunberg Berates Older Generations at UN Summit for Their Disregard of Climate Issues," September 24, 2019, accessed September 30, 2019, https://en.mercopress.com/2019/09/24/thunberg-berates-older-generations-at-un-summit-for-their-disregard-of-climate-issues.

4. Aida Chavez, "Climate Activists from Extinction Rebellion Glued Themselves to the Capitol to Disrupt House Votes," The Intercept, July 23, 2019, accessed September 21, 2019, https://theintercept.com/2019/07/23/extinction-rebellion-climate-change-capitol-hill/.

5. Meghan Bobrowsky, "Camp Fire Death Toll Rises to 86 After Man Who Suffered Third-Degree Burns Dies," *The Sacramento Bee*, August 18, 2019, accessed September 21, 2019, https://www.sacbee.com/news/california/fires/article233683422.html.

6. Tara Fowler, Emily Shapiro, and Julia Jacobo, "Hurricane Irma Strengthens to Category 5 as 2nd Storm Forms Behind It," *ABC News*, September 6, 2017, accessed February 9, 2019, https://abcnews.go.com/International/hurricane-irma-strengthens-closes-caribbean/story?id=49608171.

7. Umair Irfan, "The North Pole Just Had an Extreme Heat Wave for the 3rd Winter in a Row," *Vox*, February 28, 2018, accessed February 5, 2019, https://www.vox.com/energy-and-environment/2018/2/27/17053284/arctic-heat-wave-north-pole-climate.

8. Intergovernmental Panel on Climate Change, "Summary for Policymakers of IPCC Special Report on Global Warming of 1.5°C Approved by Governments," IPCC Press Release, October 8, 2018, accessed

March 16, 2019, https://www.ipcc.ch/site/assets/uploads/2018/11/pr
_181008_P48_spm_en.pdf.

9. Timothy Cama and Miranda Green, "Interior Chief Zinke to Leave Administration," *The Hill*, December 15, 2018, accessed January 14, 2019, https://thehill.com/policy"/energy-environment/415988-interior -secretary-ryan-zinke-steps-down.

10. IDTechEx, "Biodegradable vs. Recyclable Plastics: Which Is Better for the Environment?" *Waste&Recycling Magazine*, June 23, 2020, accessed November 6, 2020, https://www.wasterecyclingmag.ca/blog /biodegradable-vs-recyclable-plastics-which-is-better-for-the-environment/.

11. David C. Mahood, *One Green Deed Spawns Another: Tales of Inspiration on the Quest for Sustainability*, (Beverly, MA: Olive Designs, LLC, 2017), 23.

Chapter 2

1. Bill Christofferson, *The Man from Clear Lake: Earth Day Founder Senator Gaylord Nelson* (Madison, WI: University of Wisconsin Press, 2004), 307.

2. Jack Lewis, "The Spirit of the First Earth Day," *EPA Journal,* January/ February 1990, accessed February 16, 2018, https://archive.epa.gov/epa /aboutepa/spirit-first-earth-day.html.

3. Adam Rome, *The Genius of Earth Day: How a 1970 Teach-In Unexpectedly Made the First Green Generation* (New York: Hill and Wang, 2013), 22.

4. Earth Day 1970 Part 13: Conclusion, CBS News with Walter Cronkite, YouTube, https://www.youtube.com/watch?v=6HUtM_LTyIw.

5. EarthDay.org, "Two New York Times Ads, Separated by Half a Century, Call Millions to Action," February 1, 2o2o, accessed April 22, 2020, https://www.earthday.org/50-years-later-the-new-york-times-runs-another -full-page-ad-for-earth-day/.

6. Maya Wei-Haas, "Red Tide Is Devastating Florida's Sea Life. Are Humans to Blame?" *National Geographic*, August 18, 2018, accessed January 25, 2019, https://www.nationalgeographic.com/environment/article /news-longest-red-tide-wildlife-deaths-marine-life-toxins.

7. Woods Hole Oceanographic Institute, "Scientists Pinpoint How Ocean Acidification Weakens Coral Skeletons," *WHOI.edu*, January 28, 2019,

accessed June 6, 2019, https://www.whoi.edu/press-room/news-release/scientists-identify-how-ocean-acidification-weakens-coral-skeletons/.

8. Thirst Project, "Water Crisis: 663 Million People on Our Planet Lack Access to Safe, Clean Drinking Water!" n.d., accessed February 8, 2019, https://www.thirstproject.org/water-crisis/?gclid=EAIaIQobChMIgOOs6O-2s4AIVwlqGCh2N2AGdEAAYASAAEgIAIfD_BwE.

9. Anna Clark, *The Poisoned City: Flint's Water and the American Urban Tragedy* (New York: Metropolitan Books, 2018), 8.

10. Michael Moore, "Flint Poisoning Is a Racial Crime," *TIME*, January 21, 2016, accessed March 25, 2019, https://time.com/4188323/michael-moore-flint-racial-crime/.

11. Frances Moore Lappé, *Diet for a Small Planet* (New York: Ballantine Books, 1971), xlii.

12. Intergovernmental Panel on Climate Change, "Summary for Policymakers of IPCC Special Report on Global Warming of 1.5°C Approved by Governments," IPCC Press Release, October 8, 2018, accessed November 12, 2019. https://www.ipcc.ch/site/assets/uploads/2018/11/pr_181008_P48_spm_en.pdf.

13. Rebecca Hersher and Allison Aubrey, "To Slow Global Warming, U.N. Warns Agriculture Must Change," NPR. *All Things Considered*, August 8, 2019, accessed November 19, 2019, https://www.npr.org/sections/thesalt/2019/08/08/748416223/to-slow-global-warming-u-n-warns-agriculture-must-change.

14. Tad Friend, "Can a Burger Help Solve Climate Change? Eating Meat Creates Huge Environmental Costs. Impossible Foods Thinks It Has a Solution," *New Yorker Magazine*, September 30, 2019, 42.

15. David Hasemyer, John H. Cushman Jr. and Neela Banerjee, "CO2's Role in Global Warming Has Been on the Oil Industry's Radar Since the 1960s," *Inside Climate News*, April 13, 2016, accessed December 5, 2019, https://insideclimatenews.org/news/13042016/climate-change-global-warming-oil-industry-radar-1960s-exxon-api-co2-fossil-fuels/.

16. George M. Woodwell, Gordon J. MacDonald, Roger Revelle and C. David Keeling, "The Carbon Dioxide Problem:Implications for Policy in the Management of Energy and Other Resources: A Report to the Council on Environmental Quality, July 1979," *New York Times*, reprinted October 24, 2008, accessed April 3, 2019, http://graphics8.nytimes.com/packages/pdf/science/woodwellreport.pdf.

17. Nathaniel Rich, *Losing Earth: A Recent History* (New York: MCD/Farrar, Straus and Giroux, 2019), 144.

18. Patrick Hardin, "Cartoon of the Week for February 03, 1999," *Funny Times*, February 3, 1999, accessed March 26, 2019, https://funnytimes.com/19990203/.

19. Shashank Bengali, "Your Trash Is Suffocating This Indonesian Village. Here's How:" latimes.com, October 25, 2019, accessed March 16, 2020, https://www.latimes.com/world-nation/story/2019-10-25/plastic-pollution-waste-recycling-indonesia.

20. NBC News, "Mattel Issues New Massive China Toy Recall," August 13, 2007, accessed May 20, 2020, https://www.nbcnews.com/id/wbna20254745.

21. John Sutter and Lawrence Davidson, "Teen Tells Climate Negotiators They Aren't Mature Enough," CNN, December 17, 2018, accessed March 18, 2019, https://edition.cnn.com/2018/12/16/world/greta-thunberg-cop24/index.html.

Chapter 3

1. NASS, "Cattle Death Loss," USDA Library, Cornell, May 5, 2006, accessed February 19, 2019, https://downloads.usda.library.cornell.edu/usda-esmis/files/vh53wv75j/d504rp01h/8s45qc324/CattDeath-05-05-2006.pdf.

2. Shawn Cantrell (Defenders of Wildlife), interview by David Mahood, February 26, 2019.

3. Laura Zuckerman, "Controversial Idaho Hunting Contest Ends with No Wolves Killed," *Scientific American*, December 21, 2013, accessed February 28, 2019, https://www.scientificamerican.com/article/controversial-idaho-hunting-contest/.

4. "Timeline: The Fight for Northern Rocky Gray Wolves," *EarthJustice*, n.d., accessed February 15, 2019, https://earthjustice.org/features/campaigns/wolves-in-danger-timeline-milestones.

5. Christopher Weber, "Judge Upholds Protection for Gray Wolves in California," *ABC News*, January 28, 2019, accessed March 1, 2019, https://abcnews.go.com/Technology/wireStory/judge-upholds-protection-gray-wolves-california-60691554.

6. Shawn Cantrell (Defenders of Wildlife), interview by David Mahood, February 26, 2019.

7. Brodie Farquhar, "Wolf Reintroduction Changes Ecosystem in Yellowstone," *Yellowstone National Park Trips*, June 30, 2020, accessed July 16, 2020, https://www.yellowstonepark.com/things-to-do/wildlife/wolf-reintroduction-changes-ecosystem/.

8. Katherine Arberg, "Groups Challenge Trump Administration Over Gray Wolf Delisting," *Defenders of Wildlife*, January 14, 2021, accessed January 21, 2021, https://defenders.org/newsroom/groups-challenge-trump-administration-over-gray-wolf-delisting.

9. Jeff Tollefson, "Wolves Cut Yellowstone Coyote Numbers," *Timber Wolf Information Network*, n.d., accessed August 3, 2019, http://www.timber-wolfinformation.org/wolves-cut-yellowstone-coyote-numbers/.

10. U.S. Fish and Wildlife Service, "Endangered Species Act | A History of the Endangered Species Act of 1973." *FWS.gov*, last updated January 30, 2020, accessed March 6, 2020, https://www.fws.gov/endangered/laws-policies/esa-history.html.

11. Charlie Campbell, "Traditional Chinese Medical Authorities Are Unable to Stop the Booming Trade in Rare Animal Parts," *TIME*, Nov. 21, 2016, accessed January 11, 2019, https://news.yahoo.com/traditional-chinese-medical-authorities-unable-032417027.html.

12. Jes Burns, "Looking Back: The Northwest Forest Plan's New Conservation Paradigm," *Oregon Public Broadcasting*, April 16, 2015, accessed January 25, 2019, https://www.opb.org/news/article/looking-back-the-northwest-forest-plans-new-conservation-paradigm/.

13. Forest Stewardship Council, "What We Do," n.d., accessed January 26, 2019, https://us.fsc.org/en-us/what-we-do.

14. Louie Pérez, "*Good Morning Aztlán: The Words, Pictures & Songs of Louie Pérez*" (San Fernando, CA: Tía Chucha Press, 2018), 148.

Chapter 4

1. Daniel Quinn, "*Ishmael*." (New York, NY: Bantam Books, 1992).

2. Worldometers.org, n.d., accessed December 17, 2019, https://www.worldometers.info/world-population/#table-historical.

3. World Wide Fund for Nature, "Elephants." n.d., accessed December 17, 2019, https://wwf.panda.org/discover/knowledge_hub/endangered_species/elephants/?.

4. Wolfgang Lehmacher, "Wildlife Crime: A $23 Billion Trade That's Destroying Our Planet," September 28, 2016, *World Economic Forum*, accessed December 4, 2019, https://www.weforum.org/agenda/2016/09 /fighting-illegal-wildlife-and-forest-trade.

5. National Elephant Center, "Elephant Basics," n.d., accessed March, 16 2020, http://www.nationalelephantcenter.org/learn.

6. PBS, "Rhino Horn Use: Fact vs. Fiction," Nature, August 20, 2010, accessed December 19, 2019, https://www.pbs.org/wnet/nature/rhinoceros-rhino -horn-use-fact-vs-fiction/1178/.

7. NPR, "Elephant Slaughter, African Slavery and America's Pianos," *Morning Edition with Christopher Joyce*, August 18, 2014, accessed December 17, 2019, https://www.npr.org/2014/08/18/338989248 /elephant-slaughter-african-slavery-and-americas-pianos.

8. Lydia Mulvaney and Denitsa Tsekova, "America's Obsession With Beef Is Killing Leather," *Bloomberg*, August 9, 2019, accessed December 2, 2020, https://www.bloomberg.com/news/features/2019-08-09 /america-s-obsession-with-beef-is-killing-the-leather-industry.

9. Nicoletta Lanese, "Escaped Mink Could Spread the Coronavirus to Wild Animals," *Live Science*, November 30, 2020, accessed December 4, 2020, https://www.livescience.com/escaped-mink-coronavirus-infection-risk.html.

10. James Gorman, "How Mink, Like Humans, Were Slammed by the Coronavirus," *New York Times*, December 23, 2020, accessed December 28, 2020, https://www.nytimes.com/2020/12/23/science/covid-mink-animals .html.

11. Ed Stoddard, "Illegal Mining Hits Congo Gorilla Population: Conservationists," *Reuters*, April 6, 2016, accessed January 30, 2020, https:// www.reuters.com/article/us-africa-gorillas/illegal-mining-hits-congo -gorilla-population-conservationists-idUSKCN0X30T2.

12. World Wildlife Fund, "Mountain Gorilla Numbers Rise in a Central African Protected Forest," n.d., accessed January 30, 2020, https://www.worldwildlife.org/stories/mountain-gorilla-numbers -rise-in-a-central-african-protected-forest.

13. Yuval Noah Harari, *Sapiens: A Brief History of Humankind* (New York: Harper Perennial, 2015), 74.

14. Bill Hutchinson, "2 Murders Linked to Mexico"s Famed Monarch Butterfly Sanctuary," *ABC News*, February 4, 2020, accessed February 11, 2020, https://abcnews.go.com/International/double

-murder-mystery-linked-mexicos-famed-monarch-butterfly /story?id=68746901.

15. David Quammen, *The Song of the Dodo: Island Biogeography in an Age of Extinctions*" (New York: Scribner, 1996), Chapter 6.

16. Laurel Hamers, "When Bogs Burn, the Environment Takes a Hit," *Science News*, March 6, 2018, accessed December 18, 2019, https://www.sciencenews.org/article/bogs-peatlands-fire-climate-change.

17. Alexander Navarro, "World Wakes Up to Oreo's Dirty Palm Oil Secret," Greenpeace, November 23, 2018, accessed December 17, 2019, https://www.greenpeace.org/international/story/19521/the-world"-wakes-up-to-oreos-dirty-secret/.

18. Reuters Staff, "Singapore Smog Worst in Three Years as Forest Fires Rage," Reuters, September 14, 2019, accessed December 17, 2019, https://www.reuters.com/article/us-southeastasia-haze-singapore/singapore-smog-worst-in-three-years-as-forest-fires-rage-idUSKBN1VZ086.

19. Donavyn Coffey, "What Is Coral Bleaching?" *Live Science*, January 31, 2019, accessed January 4, 2020, https://www.livescience.com/64647-coral-bleaching.html.

20. NOAA, Arctic Program, "Arctic Report Card: Update for 2017," Executive Summary, November 17, 2017, accessed January 16, 2020, https://arctic.noaa.gov/Report-Card/Report-Card-2017/ArtMID/7798/ArticleID/685/Executive-Summary.

21. CBC News, "Climate Change Pushing Lobster North, Study Says," September 21, 2013, accessed September 3, 2020, https://www.cbc.ca/news/canada/nova-scotia/climate-change-pushing-lobster-north-study-says-1.1863271.

22. Yanxi Fang, "A Whale of a Problem: Japan's Whaling Policies and the International Order," *Harvard International Review*, October 23, 2019, accessed January 14, 2020, https://hir.harvard.edu/a-whale-of-a-problem-japans-whaling-policies-and-the-international-order/.

23. Erika I. Ritchie, "Lawsuit Threatened over Ship Strikes on Whales Near Ports," *Orange County Register*, March 10, 2020, accessed September 9, 2020, https://www.ocregister.com/2020/03/10/lawsuit-threatened-over-ship-strikes-on-whales-near-ports/.

24. David Abel, "New Regulations Are a Matter of Life and Extinction for Right Whales," *Boston Globe*, April 20, 2019, accessed January 15, 2020, https://www.bostonglobe.com/metro/2019/04/20/how

-federal-officials-hope-protect-endangered-right-whales-cape-cod-bay
/I57nFpeibxyuslQDeevYYO/story.html.

25. Carl Safina, *Becoming Wild: How Animal Cultures Raise Families, Create Beauty, and Achieve Peace,* (New York: Henry Holt, 2020), 36.

26. *Star Trek IV: Voyage Home,* directed by Leonard Nimoy, released November 26, 1986, Paramount Pictures, USA, film.

27. Carl Safina, *Becoming Wild: How Animal Cultures Raise Families, Create Beauty, and Achieve Peace,* (New York: Henry Holt, 2020), 327.

28. James Nestor at the Interval, "Sperm Whales Clicking You Inside Out," YouTube, April 27, 2017, accessed September 7, 2020, https://www.youtube.com/watch?v=zsDwFGz0Okg.

29. Rebecca Giggs, *Fathoms: The World in the Whale* (New York: Simon & Schuster, 2020), 164-167.

30. Ben Wolford, "Whales Are Being Killed by Noise Pollution," *Newsweek,* April 2, 2014, accessed September 4, 2020, https://www.newsweek.com/2014/04/11/whales-are-being-killed-noise-pollution-248069.html.

Chapter 5

1. Bruce Springsteen, "4th of July, Asbury Park (Sandy)" on *The Wild, the Innocent and the E Street Shuffle* album (New York: Columbia Records, 1973).

2. National Geographic, Resource Library, "Tornadoes and Climate Change," October 24, 2019, accessed June 3, 2020, https://www.nationalgeographic.org/article/tornadoes-and-climate-change/.

3. IMDb, *Jersey Shore* (2009-2012) TV Series, IMDb, n.d., accessed June 2, 2020, https://www.imdb.com/title/tt1563069/?ref_=ttfc_fc_tt.

4. *Saving Private Ryan,* directed by Steven Spielberg, released July 24, 1998, Dreamworks Pictures, Paramount Pictures, USA, film.

5. Sarah Gibbens, "Hurricane Sandy, Explained," *National Geographic,* February 11, 2019, accessed May 13, 2020, https://www.nationalgeographic.com/environment/article/hurricane-sandy.

6. John P. Rafferty, "Superstorm Sandy," *Encyclopedia Britannica,* updated, March 2, 2021, accessed June 1, 2020, https://www.britannica.com/event/Superstorm-Sandy.

7. Jen Francis, *"Arctic Meltdown and Unruly Tropical Storms: How are they connected?"* (webinar), Woods Hole Research Center, via Zoom Meetings May 20, 2020.

8. Jean Mickle, "Ortley Beach Was Sandy's 'Ground Zero,' "Asbury Park Press, October 24, 2017, accessed June 3, 2020, https://www .app.com/story/news/local/ocean-county/sandy-recovery/2017/10/24 /ortley-beach-superstorm-sandys-ground-zero-5-years-ago/783933001/.

9. Ross Toro, "Hurricane Sandy's Impact (Infographic)" *Live Science*, October 29, 2013, accessed December 11, 2019, https://www.livescience .com/40774-hurricane-sandy-s-impact-infographic.html.

10. Miles Grant, "Polar vortex puts Woodwell senior scientist in national spotlight," *Woodwell Climate Research Center*, February 19, 2019, accessed May 20, 2020, https://www.woodwellclimate.org /polar-vortex-puts-woodwell-senior-scientist-in-national-spotlight/.

11. Erin O'Neill, "More than $1B Spent Replenishing N.J. Beaches over Past 30 Years," *New Jersey Advance Media*, January 17, 2019, accessed December 13, 2019, https://www.nj.com/news/2015/10/jersey_shore _beach_replenishment_cost_pegged_at_1b_over_past_30_years.html.

12. Mike Scott, "Insurers Will Be Hard-Hit By Climate Change But They're Not Investing In The Low-Carbon Economy," *Forbes*, May 31, 2018, accessed December 13, 2019, https://www.forbes.com/sites/mikescott /2018/05/31/insurers-in-the-front-line-of-the-fight-against-climate-change -shoot-themselves-in-the-foot/?sh=292c015340fa.

13. Laura Sullivan, "Business Of Disaster: Insurance Firms Profited $400 Million After Sandy," *NPR*, *All Things Considered*, May 24, 2016, accessed December 11, 2019, https://www.npr.org/2016/05/24/478868270 /business-of-disaster-insurance-firms-profited-400-million-after-sandy.

14. "The Future of Sea Level Rise: Sea Level Rise Is Speeding Up," Sea Level Rise.org, n.d., accessed December 16, 2019, https://sealevelrise .org/forecast/.

15. "The Backstreet Phantom of Rock," *TIME*, October 27, 1975, accessed December 16, 2019, http://content.time.com/time/subscriber/article /0,33009,913583,00.html.

16. Chris Glorioso and Tom Burke, "I-Team Tracks Sandy Relief Money from 12-12-12 Concert," *NBC New York*, February 26, 2013, accessed December 9, 2019, https://www.nbcnewyork.com/news/local /sandy-concert-12-12-relief-funds-grants-robin-hood/2075223/.

17. "Hurricane Sandy's Impact, By The Numbers (INFOGRAPHIC)," *Huffington Post*, October 29, 2013, accessed August 16, 2019, https://www.huffpost.com/entry/hurricane-sandy-impact-infographic_n_4171243.

18. Stephanie Vierra, "Green Building Standards and Certification Systems," Whole Building Design Guide, August 25, 2019, accessed June 18, 2020, https://www.wbdg.org/resources/green-building-standards-and-certification-systems.

Chapter 6

1. Michael Wines, "Monarch Migration Plunges to Lowest Level in Decades," *New York Times*, March 13, 2013, accessed January 9, 2014, https://www.nytimes.com/2013/03/14/science/earth/monarch-migration-plunges-to-lowest-level-in-decades.html.

2. Tierra Curry and George Kimbrell, "Eastern Monarch Butterfly Population," *Center for Biological Diversity*, January 30, 2019, accessed July 7, 2020, https://www.biologicaldiversity.org/news/press_releases/2019/monarch-butterfly-01-30-2019.php.

3. Yale School of the Environment, "Western U.S. Monarch Butterfly Population Declined Sharply in 2018," *Yale Environment 360*, January 9, 2019, accessed July 7, 2020, https://e360.yale.edu/digest/western-u-s-monarch-butterfly-population-declined-sharply-in-2018.

4. Charles M. Benbrook, "Trends in Glyphosate Herbicide Use in the United States and Globally," Environmental Sciences Europe, February 2, 2016, accessed July 16, 2020, https://enveurope.springeropen.com/articles/10.1186/s12302-016-0070-0.

5. Carey Gillam, *Whitewash: The Story of a Weed Killer, Cancer, and the Corruption of Science* (Washington, DC: Island Press, 2017), 84.

6. Katy Moncivais, "Roundup Lawsuit," Consumer Safety, n.d., accessed July 16, 2020, https://www.consumersafety.org/product-lawsuits/roundup/.

7. Uwe Hessler, "What's Driving Europe's Stance on Glyphosate," Deutsche Welle, June 25, 2020, accessed July 16, 2020, https://www.dw.com/en/whats-driving-europes-stance-on-glyphosate/a-53924882.

8. Sudeep Chakravarty, "World's Top 10 Agrochemical Companies: Industry Forecast and Trends," Market Research Reports, September 25, 2019, accessed July 20, 2020. https://www.marketresearchreports.com/blog/2019/09/25/worlds-top-10-agrochemical-companies-industry-forecast-and-trends.

9. Christina Sarich, "Monsanto Sues Farmers for 16 Straight Years over GMOs, NEVER Loses," Natural Society, August 29, 2014, accessed June 1, 2020, https://naturalsociety.com/monsanto -sued-farmers-16-years-gmos-never-lost/.

10. "Supreme Court Hands Monsanto Victory over Farmers on GMO Seed Patents, Ability to Sue," *RT News,* January 13, 2014, accessed June 2, 2020, https://www.rt.com/usa/monsanto-patents-sue-farmers-547/.

11. Green America, "GMOs Don't Feed the World," n.d., accessed June 1, 2020, https://www.greenamerica.org/gmos-stop-ge-wheat/genetic -engineering-gmos/gmos-dont-feed-world.

12. Danny Hakim, "Doubts About the Promised Bounty of Genetically Modified Crops," *New York Times,* October 29, 2016, accessed June 1, 2020, https://www.nytimes.com/2016/10/30/business/gmo-promise-falls-short .html?ref=todayspaper&_r=0.

13. Michele Colopy, "Pollinator Habitat Is Disappearing At Rates Usually Reserved For Descriptions Of Amazon Rain Deforestation," Pollinator-Pathway, April 8, 2015, accessed June 1, 2020, https://www .pollinator-pathway.org/roundup.

14. Deidre Imus, "New Study Finds Organic Foods Are Healthier than Conventionally Grown Foods," Fox News, October 24, 2015, accessed June 3, 2020, https://www.foxnews.com/health/new-study-finds-organic-foods -are-healthier-than-conventionally-grown-foods.

15. Ron Cook, "World Beef Production: Ranking of Countries," Beef2Live, April 6, 2021, accessed April 8, 2021, https://beef2live.com /story-world-beef-production-ranking-countries-0-106885.

16. Genetic Roulette: The Gamble of our Lives, a film by Jeffery M. Smith, released in August, 2012, Institute for Responsible Technology, USA, film.

17. Kathleen Doheny, "Rise in Chronic Childhood Health Problems," WebMD, February 16, 2010, accessed June 25, 2020, https://www.webmd.com /children/news/20100216/rise-in-childhood-health-problems#1.

18. Frances Moore Lappé, Diet for a Small Planet (New York: Ballantine Books, 1971).

19. Science in Society, "Independent Scientists Manifesto on Glyphosate," August 6, 2015, accessed June 5, 2020, http://www.i-sis.org.uk/Independent _Scientists_Manifesto_on_Glyphosate.php.

20. *The Devil We Know,* directed by Stephanie Soechtig and Jeremy Seifert, released January 21, 2018, Cinetic Media, USA, film.

21. *Dark Waters*, directed by Todd Haynes, released December 6, 2019, Focus Features, USA, film.

22. Environmental Working Group, "The 'Forever Chemicals' in 99% of Americans," n.d., accessed June 12, 2020, https://www.ewg.org/pfaschemicals /what-are-forever-chemicals.html.

23. U.S. Environmental Protection Agency, "EPA Takes Final Step in Phase-out of Leaded Gasoline," January 29, 1996, accessed June 10, 2020, https://archive.epa.gov/epa/aboutepa/epa-takes-final-step-phaseout-leaded -gasoline.html.

24. Kirk Semple and Paulina Villegas, "Mexican Butterfly Conservationist Is Found Dead, Two Weeks After Vanishing," *New York Times*, January 20, 2020, accessed January 31, 2020, https://www.nytimes.com/2020/01/29 /world/mexico-butterfly-dead.html.

25. AG America Lending, "Power of 10: Top 10 Pollinators in Agriculture," July 10, 2017, accessed July 11, 2020, https://agamerica.com/blog /top-10-pollinators-in-agriculture/.

26. U.S. Environmental Protection Agency, "EPA Releases Proposed Interim Decisions for Neonicotinoids," January 30, 2020, accessed June 22, 2020, https://www.epa.gov/pesticides/epa-releases-proposed -interim-decisions-neonicotinoids.

27. David Wolfe, "How Climate Change Affects the Monarch Butterfly, and What We Can Do about It," Environmental Defense Fund, May 26, 2016, accessed June 2, 2020, https://www.edf.org/blog/2016/05/26/how-climate -change-affects-monarch-butterfly-and-what-we-can-do-about-it.

28. Keith Randall, "Monarch Butterfly Numbers Down Again," *Texas A&M Today*, March 19, 2020, accessed July 7, 2020, https://today.tamu.edu /2020/03/19/monarch-butterfly-numbers-down-again/.

Chapter 7

1. NEWH, Vision and Mission, n.d. accessed February 4, 2020, https://newh .org/about-us/vision-and-mission/.

2. Climate Reality Project, "The Climate Denial Machine: How the Fossil Fuel Industry Blocks Climate Action," September 5, 2019, accessed February 3, 2020, https://www.climaterealityproject.org/blog /climate-denial-machine-how-fossil-fuel-industry-blocks-climate-action.

3. Shannon Hall, "Exxon Knew about Climate Change almost 40 years ago," Scientific American, October 26, 2015, accessed February 19, 2020, https://www.scientificamerican.com/article /exxon-knew-about-climate-change-almost-40-years-ago/.

4. Neela Banerjee, "Members of an American Petroleum Institute Task Force on CO2 Included Scientists from Nearly Every Major Oil Company, Including Exxon, Texaco and Shell," *Inside Climate News*, December 22, 2015, accessed March 13, 2020, https://insideclimatenews.org/news /22122015/exxon-mobil-oil-industry-peers-knew-about-climate-change -dangers-1970s-american-petroleum-institute-api-shell-chevron-texaco/.

5. Nathaniel Rich, *Losing Earth: A Recent History* (New York: MCD/Farrar, Straus and Giroux, 2019), 48.

6. Michael E. Mann and Tom Toles, *The Madhouse Effect: How Climate Change Denial Is Threatening Our Planet, Destroying Our Politics, and Driving Us Crazy* (New York: Columbia University Press, 2016), 6.

7. Michael E. Mann, *The New Climate War: The Fight to Take Back Our Planet* (New York: Hachette Book Group, 2021), 87.

8. Lydia Saad, "Global Warming Concern at Three-Decade High in U.S.," Gallup, March 14, 2017, accessed February 5, 2020, https://news.gallup .com/poll/206030/global-warming-concern-three-decade-high.aspx.

9. DARA, "Climate Vulnerability Monitor, 2nd Edition," daraint.org, accessed February 8, 2020, https://daraint.org/wp-content/uploads/2012/09 /EXECUTIVE-AND-TECHNICAL-SUMMARY.pdf.

10. World Health Organization, "WHO Global Programme on Climate Change & Health," n.d., accessed February 19, 2020, https://www.who .int/globalchange/mediacentre/news/global-programme/en/.

11. Noah Kaufman, John Larsen, Peter Marsters, Hannah Kolus, and Shashank Mohan, "An Assessment of the Energy Innovation and Carbon Dividend Act," Columbia Center on Global Energy Policy, November 6, 2019, accessed February 22, 2020, https://www.energypolicy.columbia.edu /research/report/assessment-energy-innovation-and-carbon-dividend-act.

Chapter 8

1. Jennifer Ann Demoss, "It's a Banner Year for Sea Turtles along the North Carolina Coast. What's Behind It?" *News & Observer*, August 8, 2019, accessed March 18, 2020, https://www.newsobserver.com/news/local /article233441247.html.

2. Mike Shutak and Anna Harvey," Rescue Facilities Care for Cold-Stunned Sea Turtles," *Carteret County News-Times*, January 3, 2013, accessed April 13, 2020, https://www.carolinacoastonline.com/news_times/news /article_69f2a5da-5a76-11e2-8aca-0019bb2963f4.html.

3. Turtle Conservancy, "About Us," n.d., accessed May, 5, 2020, https:// www.turtleconservancy.org/about.

4. ABC News, " 'Great Pacific Garbage Patch' Is Massive Floating Island of Plastic, Now 3 Times the Size of France," *ABC News*, March 23, 2018, accessed May 9, 2020, https://abcnews.go.com/International/great-pacific -garbage-patch-massive-floating-island-plastic/story?id=53962147.

5. Matthew Robinson, "Microplastics Found in Gut of Every Sea Turtle in New Study," CNN, December 5, 2018, accessed May 5, 2020, https:// www.cnn.com/2018/12/05/world/microplastic-pollution-turtles-study -intl-scli/index.html.

6. Theodore Seuss Geisel, *The Lorax* (New York: Random House, 1971).

7. Ibid.

8. Simon Reddy, "Plastic Pollution Affects Sea Life Throughout the Ocean," Pew Trusts, September 24, 2018, accessed February 9, 2020, https://www.pewtrusts.org/en/research-and-analysis/articles/2018/09/24 /plastic-pollution-affects-sea-life-throughout-the-ocean.

9. David C. Mahood, *One Green Deed Spawns Another: Tales of Inspiration on the Quest for Sustainability* (Beverly, MA: Olive Designs, LLC, 2017), 53.

10. PBS, "Plastic Wars," *Frontline*, April 1, 2020, WGBH, film.

11. U.S. Environmental Protection Agency, "Facts and Figures about Materials, Waste and Recycling," n.d., accessed March 21, 2020, https: //www.epa.gov/facts-and-figures-about-materials-waste-and-recycling /plastics-material-specific-data#PlasticsTableandGraph.

12. Isobel Whitcomb, "How Much Plastic Actually Gets Recycled?" *Live Science*, March 7, 2020, accessed May 6, 2020, https://www.livescience. com/how-much-plastic-recycling.html.

13. Daniel Brockett, "How Plastic Is Made from Natural Gas," Penn State Extension, January 17, 2017, accessed April 19, 2020, https://extension .psu.edu/how-plastic-is-made-from-natural-gas.

14. Joseph Winters, "It's Official: Reusables Are Safe during COVID-19," Grist, June 26, 2020, accessed August 11, 2020, https://grist.org/climate /its-official-reusables-are-safe-during-covid-19/.

15. Biomimicry 3.8, "What is Biomimicry?" n.d., accessed June 21, 2020, https://biomimicry.net/what-is-biomimicry/.

16. Biomimicry Institute, "Solutions to Global Challenges Are All around Us," n.d., accessed June 21, 2020, https://biomimicry.org/biomimicry-examples/.

17. Mark Dorfman, "Biology Inspires a Plastics (R)Evolution," Biomimicry 3.8 (blog), April 20, 2020, accessed April 24, 2020, https://synapse.bio/blog /biology-inspires-a-plastics-revolution?fbclid=IwAR05p8eOJL-gXw5RG2XRC0BQS09dBjx4GcgmmD5Y0fFViYg81a57jLNIgxxY.

18. Amy Woodyatt, "Hermit Crabs Are Confusing Plastic for Shells and It's Killing Them," CNN, December 15, 2019, accessed February 7, 2021, https://www.cnn.com/2019/12/05/world/hermit-crabs-plastic-pollution -intl-scli-scn/index.html.

19. Peter C.H. Pritchard, *Encyclopedia of Turtles* (Neptune, New Jersey: T.H.F Productions, 1979).

20. Kevin Spear, "Turtle Conservationist Peter Pritchard Dies at 76," *Orlando Sentinel*, February 26, 2020, accessed February 28, 2020, https://www .orlandosentinel.com/news/environment/os-ne-peter-pritchard-turtle-fame -obit-20200226-t27wdr4vandpzm33iik74jxybi-story.html.

21. Mike O'Neill, email conversation with author, April 11, 2020.

Chapter 9

1. Alfred, Lord Tennyson, ed. Stanley Appelbaum, *The Charge of the Light Brigade and Other Poems* (Mineola, NY: Dover Publications, 1992), 52.

2. U.S. Energy Information Administration, "Frequently Asked Questions (FAQS), Which States Produce the Most Coal?" n.d., accessed November 3, 2020, https://www.eia.gov/tools/faqs/faq.php?id=69&t=2.

3. Environmental Defense Fund, "Clean Energy Is Building a New American Workforce," January, 2018, accessed January 12, 2020, https://www.edf .org/energy/clean-energy-jobs.

4. Citizens' Climate Lobby, "Climate Solutions Caucus," n.d., accessed March 3, 2017, https://citizensclimatelobby.org/climate-solutions-caucus/.

5. *An Inconvenient Truth*, directed by Davis Guggenheim, produced by Lawrence Bender Productions and Participant Media, Paramount Vantage, 2006, film.

6. James Hansen, *Storms of My Grandchildren: The Truth About the Coming Catastrophe and Our Last Chance to Save Humanity* (New York: Bloomsbury, 2009), 110.

7. Ibid, 66.

8. Albert Gore Jr., *Earth in the Balance: Ecology and the Human Spirit* (New York: Plume, 1973), 5.

9. Nathaniel Rich, *Losing Earth: A Recent History* (New York: MCD/Farrar, Straus and Giroux, 2019), 76-77.

10. "A Conversation with Bill McKibben," *Grist*, December 21, 2010, accessed October 14, 2020, https://grist.org/article/2010-12-21-a-conversation-with-bill-mckibben/.

11. Wal van Lierop, "Yes, Fossil Fuel Subsidies Are Real, Destructive And Protected By Lobbying," *Forbes*, December 6, 2019, accessed November 11, 2020, https://www.forbes.com/sites/walvanlierop/2019/12/06/yes-fossil-fuel-subsidies-are-real-destructive-and-protected-by-lobbying/?sh=7749b087417e.

12. Paul Hawken, Amory Lovins, and L. Hunter Lovins, *Natural Capitalism: Creating the Next Industrial Revolution* (Boston: Back Bay Books, 2000).

13. Richard Pallardy, "Deepwater Horizon Oil Spill," *Encyclopedia Britannica*, July 9, 2010, accessed November 11, 2020, https://www.britannica.com/event/Deepwater-Horizon-oil-spill.

14. David Suzuki, "Carbon pricing is an important tool to tackle climate change," David Suzuki Foundation, June 4, 2018, accessed December 2, 2020, https://davidsuzuki.org/story/carbon-pricing-is-an-important-tool-to-tackle-climate-change/.

15. Citizens' Climate Lobby, "The Basics of Carbon Fee and Dividend: How Carbon Fees and Dividends Work," n.d., accessed January 7, 2020, https://citizensclimatelobby.org/basics-carbon-fee-dividend/.

16. *An Inconvenient Truth*, directed by Davis Guggenheim, produced by Lawrence Bender Productions and Participant Media, Paramount Vantage, 2006, film.

17. Bill McKibben, "Global Warming's Terrifying New Math," *Rolling Stone Magazine*, July 19, 2012, accessed December 29, 2020, https://www.rollingstone.com/politics/politics-news/global-warmings-terrifying-new-math-188550/.

18. David C. Mahood, *One Green Deed Spawns Another: Tales of Inspiration on the Quest for Sustainability* (Beverly, MA: Olive Designs, LLC, 2017), 8.

19. Laurie Stone, "Green Jobs for a Brighter Post-Pandemic Future," *Energy Transition Magazine*, May 2020, accessed November 11, 2020, https://theenergytransition.org/article/green-jobs-for-a-post-pandemic-bright-future/.

20. Zeke Hausfather, "State of the Climate: 2020 on Course to be Warmest Year on Record," Carbon Brief, October 23, 2020, accessed November 21, 2020, https://www.carbonbrief.org/state-of-the-climate-2020-on-course-to-be-warmest-year-on-record.

Chapter 10

1. *The Reluctant Radical*, directed by Lindsey Grayzel, co-directed by Deia Schlosberg, produced by Goodwin-Grayzel Productions, 2018, film.

2. David C. Mahood, *One Green Deed Spawns Another: Tales of Inspiration on the Quest for Sustainability* (Beverly, MA: Olive Designs, LLC, 2017), 186.

3. Danielle Torrent Tucker, "In a Warming World, Cape Town's 'Day Zero' Drought Won't be an Anomaly, Stanford Researcher Says," Stanford University, November 9, 2020, accessed December 21, 2020, https://news.stanford.edu/2020/11/09/cape-towns-day-zero-drought-sign-things-come/.

4. Salem Film Fest, n.d., accessed December 19, 2019, https://www.salemfilmfest.com/.

5. Erin Ross, "Protesters Arrested at Zenith Oil Terminal," OPB, April 22, 2019, accessed May 7, 2020, https://www.opb.org/news/article/portland-protesters-arrested-zenith-energy-oil-terminal-railroad/.

6. Robin McKie, "Green Ships in Deadly Duel with Whalers," *The Guardian*, January 8, 2013, accessed May 3, 2020, https://www.theguardian.com/environment/2008/jan/13/whaling.antarctica.

7. Lee Rowland and Vera Eidelman, "Where Protests Flourish, Anti-Protest Bills Follow," *ACLU*, February 17, 2017, accessed May 5, 2020, https://www.aclu.org/blog/free-speech/rights-protesters/where-protests-flourish-anti-protest-bills-follow?redirect=blog/speak-freely/where-protests-flourish-anti-protest-bills-follow.

8. Kerri Bartlett, "Protesters Who Camp on Tennessee State Property Could Lose Right to Vote under New Law," *USA Today*, August 22, 2020, https://www.usatoday.com/story/news/politics/2020/08/22/tennessee-law-protesters-could-lose-right-vote-camping-on-state-land/3420661001/.

9. Christian Alexander, "Cape Town's 'Day Zero' Water Crisis, One Year Later," *Bloomberg*, April 12, 2019, accessed December 12, 2019, https://www.bloomberg.com/news/articles/2019-04-12 /looking-back-on-cape-town-s-drought-and-day-zero.

10. Kevin O'Reilly, "Critical Thinking about Climate Change," PowerPoint presentation, Green Energy Forum, Beverly High School, Beverly, MA, January 6, 2020.

11. Theodore Seuss Geisel, *The Lorax* (New York: Random House, 1971).

12. Solar Now Inc. John W. Coleman Greenergy Park, n.d., accessed December 14, 2019, http://www.solarnow.org/home.html.

13. U.S. Energy Information Administration, "U.S. Energy Facts Explained," May 7, 2020, accessed August 19, 2020, https://www.eia.gov /energyexplained/us-energy-facts/.

14. CNN, "Hurricane Katrina Statistics Fast Facts," August 12, 2020, accessed August 21, 2020, https://www.cnn.com/2013/08/23/us/hurricane -katrina-statistics-fast-facts/index.html.

Chapter 11

1. Finis Dunaway, "The 'Crying Indian' Ad That Fooled the Environmental Movement," *Chicago Tribune*, November 21, 2017, accessed March 26, 2019, https://www.chicagotribune.com/opinion/commentary/ct-perspec -indian-crying-environment-ads-pollution-1123-20171113-story.html.

2. U.S. Census Bureau, "Selected Population Profile in the United States," 2017, accessed February 16, 2019, https://data.census.gov/cedsc i/table?q=ACSSPP1Y2017.S0201&t=006%20-%20American%20 Indian%20and%20Alaska%20Native%20alone%20%28300,%20A01- Z99%29&tid=ACSSPP1Y2017.S0201.

3. Centers for Disease Control and Prevention, "American Indian & Alaska Native Populations," n.d., accessed March 3, 2019, https://web.archive. org/web/20131122110259/http://www.cdc.gov/minorityhealth/popula- tions/REMP/aian.html#10.

4. National Indian Gaming Commission, Mavis Harris, "2018 Indian Gaming Revenues of $33.7 Billion Show a 4.1% Increase," September 12, 2019, accessed February 16, 2020, https://www.nigc.gov/news /detail/2018-indian-gaming-revenues-of-33.7-billion-show-a-4.1-increase.

5. Marguerite L. Sagna and Sanhita Gupta and Clare Torres, "Health Status Profile of American Indians in Arizona," Arizona Department of Health Services, February 2017, accessed March 17, 2019, https://pub.azdhs .gov/health-stats/report/hspam/2015/indian2015.pdf.

6. Dee Brown, *Bury My Heart at Wounded Knee: An Indian History of the American West* (New York: Henry Holt, 1970).

7. David Grann, *Killers of the Flower Moon: The Osage Murders and the Birth of the FBI* (New York: Doubleday, 2017).

8. Dylan Matthews, "9 Reasons Christopher Columbus Was a Murderer, Tyrant, and Scoundrel," VOX, October 12, 2015, accessed May 16, 2019, https://www.vox.com/2014/10/13/6957875/christopher -columbus-murderer-tyrant-scoundrel.

9. National Park Service, "The Dawes Act," n.d., accessed May 16, 2019, https://www.nps.gov/articles/000/dawes-act.htm.

10. Nick Estes and Alleen Brown, "Where Are the Indigenous Children Who Never Came Home?" *High Country News*, September 25, 2018, accessed May 23, 2019, https://www.hcn.org/articles/tribal-affairs-where-are-the -indigenous-children-that-never-came-home-carlisle-indian-school -nations-want-answers.

11. Lois Crozier-Hogle and Darryl Babe Wilson, *Surviving in Two Worlds: Contemporary Native Voices*, ed. by Jay Leibold. (Austin, TX: University of Texas Press, 1997).

12. David C. Mahood, *One Green Deed Spawns Another: Tales of Inspiration on the Quest for Sustainability* (Beverly, MA: Olive Designs, LLC, 2017), 110.

13. Roxanne Dunbar-Ortiz, "Land Claims: An Indigenous People's History of the United States," In These Times, September 12, 2015, accessed May 22, 2019, https://inthesetimes.com/article/land-claims -an-indigenous-peoples-history-of-the-united-states.

14. Alex White Plume, conversation with author, April 19. 2019.

15. Peter Matthiessen, *In the Spirit of Crazy Horse* (New York: Penguin Books, 1992), Chapter 6.

16. Leonard Peltier, *Prison Writings: My Life is My Sun Dance,* ed. Harvey Arden, (New York: St. Martin's, 1999), 129-133.

17. Alex White Plume, conversation with author, April 19. 2019.

18. Chet Brokaw, "Lakota Hemp Farmer Left Broke by DEA," *Lincoln Journal Star*, July 13, 2007, accessed December 3, 2020, https://journalstar.com /business/lakota-hemp-farmer-left-broke-by-dea/article_4588e603-e0ce -5beb-ae93-0afb463926ca.html.

19. Bill Weinberg, "Standing Up in Lakota Country," Project CBD, November 21, 2020, accessed February 17, 2021, https://www.projectcbd.org/culture /standing-lakota-country.

20. Michael Astor, "Debra White Plume, Defender of Her Tribe, Is Dead at 66," *New York Times*, November 27, 2020, accessed December 10, 2020, https://www.nytimes.com/2020/11/27/us/debra-white-dead.html.

21. "U.S. Department of the Interior, "Tribal Economic Development Principles-at-a-Glance Series," n.d., accessed February 17, 2021, https://www .bia.gov/sites/bia.gov/files/assets/as-ia/ieed/ieed/pdf/Primer_Native _American_HUBZones_Final_508_Compliant.pdf.

Chapter 12

1. Wayne Mahood, aka. Dad, *A Strenuous Day* (Geneseo, NY: Milne Library, State University of Geneseo, 2015).

SELECT BIBLIOGRAPHY

〽

Adams, David Wallace. *Education for Extinction: American Indians and the Boarding School Experience, 1875–1928*. Lawrence, KS: University Press of Kansas, 1995.

Anderson, Ray C., and Robin White. *Confessions of a Radical Industrialist: People, Profits, Purpose—Doing Business by Respecting Earth*. New York: St. Martin's Press, 2009.

Attenborough, David. *A Life on Our Planet: My Witness Statement and A Vision for the Future*. New York: Grand Central, 2020.

Benyus, Janine. *Biomimicry: Innovation Inspired by Nature*. New York: Quill, 1998.

Berry, Wendell. *Window Poems*. Emeryville, CA: Shoemaker & Hoard, 2007.

Bonney, Robert John, *Hide*. Published by the author, 2016.

Brown, Dee. *Bury My Heart at Wounded Knee: An Indian History of the American West*. New York: Henry Holt, 1970.

Chamberlain, Kathleen P. *Victorio: Apache Warrior and Chief*. Norman, OK: University of Oklahoma Press, 2007.

Christofferson, Bill. *The Man from Clear Lake: Earth Day Founder Senator Gaylord Nelson*. Madison, WI: University of Wisconsin Press, 2004.

Clark, Anna. *The Poisoned City: Flint's Water and The American Urban Tragedy*. New York: Metropolitan Books, 2018.

Coleman, Elizabeth J. ed. *Here: Poems for the Planet*, Port Townsend, WA: Copper Canyon, 2019.

Crozier-Hogle, Lois, and Darryl Babe Wilson. *Surviving in Two Worlds: Contemporary Native Voices*, ed. Jay Leibold. Austin, TX: University of Texas Press, 1997.

Geisel, Theodore Seuss. *The Lorax*. New York: Random House, 1971.

Giggs, Rebecca. *Fathoms: The World in the Whale*. New York: Simon & Schuster, 2020.

Gillam, Carey. *Whitewash: The Story of a Weed Killer, Cancer, and the Corruption of Science.* Washington, DC: Island Press, 2017.

Goodall, Jane, and Phillip Berman. *Reason for Hope: A Spiritual Journey.* New York: Warner Books, 1999.

Goodrich, David. *A Hole in the Wind: A Climate Scientist's Bicycle Journey Across the United States.* New York: Pegasus Books, 2018.

Gore, Albert Jr. *Earth in the Balance: Ecology and the Human Spirit.* New York: Plume, 1993.

Grann, David. *Killers of the Flower Moon: The Osage Murders and the Birth of the FBI.* New York: Doubleday, 2017.

Greenbaum, Eli. *Emerald Labyrinth: A Scientist's Adventures in the Jungles of the Congo.* Lebanon, NH: ForeEdge, 2018.

Hansen, James. *Storms of My Grandchildren: The Truth about the Coming Catastrophe and Our Last Chance to Save Humanity. New York: Bloomsbury, 2009.*

Harari, Yuval Noah. Sapiens: A Brief History of Humankind. New York: Harper Perennial, 2015.

Hawken, Paul, ed. *Drawdown: The Most Comprehensive Plan Ever Proposed to Reverse Global Warming.* New York: Penguin Books, 2017.

Hawken, Paul, Amory Lovins, and L. Hunter Lovins. *Natural Capitalism: Creating the Next Industrial Revolution.* Boston: Back Bay Books, 2000.

Healey, Jack. *Create Your Future: A Memoir by John G. (Jack) Healey.* Los Angeles: Snail Press, 2015.

Josephy, Alvin M Jr. *500 Nations: An Illustrated History of North American Indians.* New York: Alfred A. Knopf, 1994.

Klein, Naomi. *No Is Not Enough: Resisting Trump's Shock Politics and Winning the World We Need.* Chicago, IL: Haymarket Books, 2017.

Kolbert, Elizabeth. *The Sixth Extinction: An Unnatural History.* New York: Picador, 2014.

Lappé, Frances Moore. *Diet for a Small Planet.* New York: Ballantine Books, 1971.

Lappé, Frances Moore, and Adam Eichen. *Daring Democracy: Igniting Power, Meaning, and Connection for the America We Want.* Boston: Beacon, 2017.

Louv, Richard. *Last Child in the Woods: Saving Our Children from Nature Deficit Disorder.* New York: Workman, 2005.

Mahood, David C. *One Green Deed Spawns Another: Tales of Inspiration on the Quest for Sustainability.* Beverly, MA: Olive Designs, LLC, 2017.

Mahood, Wayne. *A Strenuous Day.* Geneseo: Milne Library, State University of Geneseo, 2015.

Mann, Michael E. *The New Climate War: The Fight to Take Back Our Planet.* New York: Hachette, 2021.

Mann, Michael E., and Tom Toles. *The Madhouse Effect: How Climate Change Denial Is Threatening Our Planet, Destroying Our Politics, and Driving Us Crazy.* New York: Columbia University Press, 2016.

Masson, Jeffery Moussaieff, and Susan McCarthy. *When Elephants Weep: The Emotional Lives of Animals.* New York: Delta, 1995.

Matthiessen, Peter. *In the Spirit of Crazy Horse.* New York: Penguin Books, 1992.

McKibben, Bill. *Falter: Has the Human Game Begun to Play Itself Out?* New York: Holt, 2020.

Oreskes, Naomi, and Erik M. Conway, *Merchants of Doubt: How a Handful of Scientists Obscured the Truth on Issues from Tobacco Smoke to Global Warming.* New York: Bloomsbury, 2010.

Peltier, Leonard. *Prison Writings: My Life Is My Sun Dance*, ed. Harvey Arden. New York: St. Martin's Griffin, 1999.

Pérez, Louie. *Good Morning, Aztlán: The Words, Pictures & Songs of Louie Pérez.* San Fernando, CA: Tía Chucha Press, 2018.

Powers, Richard. *The Overstory: A Novel.* New York: W.W. Norton, 2018.

Pritchard, Peter C.H. *Encyclopedia of Turtles.* Neptune, New Jersey: T.F.H. Publications, 1979.

Quammen, David. *The Song of the Dodo: Island Biogeography in an Age of Extinctions.* New York: Scribner, 1996.

Quinn, Daniel. *Ishmael.* New York: Bantam Books, 1992.

Rich, Nathaniel. *Losing Earth: A Recent History.* New York: MCD/Farrar, Straus and Giroux, 2019.

Rome, Adam. *The Genius of Earth Day: How a 1970 Teach-In Unexpectedly Made the First Green Generation.* New York: Hill and Wang, 2013.

Romm, Joseph. *Climate Change: What Everyone Needs to Know.* New York: Oxford University Press, 2018.

Rosen, Kenneth, ed. *Voices of the Rainbow: Contemporary Poetry by American Indians*. New York: Viking, 1975.

Safina, Carl. *Becoming Wild: How Animal Cultures Raise Families, Create Beauty, and Achieve Peace*. New York: Henry Holt, 2020.

———. *Voyage of the Turtle: In Pursuit of the Earth's Last Dinosaur*. New York: Owl Books, 2007.

Stevenson, Bryan. *Just Mercy: A Story of Justice and Redemption*. New York: Spiegel & Grau, 2019.

Wilson, Edward O. *Genesis: The Deep Origin of Societies*. New York: Liveright, 2019.

———. *Naturalist*. Washington, D.C.: Island Press, 1994.

Wood, Denis. *Five Billion Years of Global Change: A History of the Land*. New York: Guilford, 2004.

Woodwell, George M. *A World to Live In: An Ecologist's Vision for a Plundered Planet*. Cambridge, MA: MIT Press, 2016.

.